Cloud Computing-
An Introduction

by

A. Sindhu Devi

B.E.,MBA.,M.TECH.,B.SC.,B.ED.

Double9 BOOKS

Cloud Computing-An Introduction

by A. Sindhu Devi

ISBN: 978-93-63055-32-2

Published by

DOUBLE 9 BOOKS
2/13-B, Ansari Road
Daryaganj, New Delhi – 110002
info@double9books.com
www.double9books.com
Tel. 011-40042856

ABOUT THE AUTHOR

The author of this book is Mrs. A. Sindhu Devi. She has done her schooling in various parts of Tamilnadu. She has completed her undergraduate degree in Electronics and Communication in First class under Anna University affiliated college. She has completed her MBA degree under Alagappa University with First class After her marriage she took a break in her studies and completed B.Sc in Mathematics B.Ed in Mathematics. She joined M.Tech in Computer science Engineering in BIHER, Chennai. She had worked in matriculation and CBS schools as Science Computer Science and Mathematics teacher. She had published papers on Cloud Computing and Network security on journals. Due to her interest in Computer science and her kids and husband's well wish she has been working on to write a book on Cloud Computing. Her kids and husband has been her backbone for her academic journey and work journey.The major reason for her author journey is her husband Mr. Venkadesh and her two kids Selvi. Vaishnavi Devi and Selvi. Saindhavi Devi.

CONTENTS

UNIT I
UNDERSTANDING CLOUD COMPUTING

Cloud Computing

Cloud computing is a general term for the delivery of hosted computing services and IT resources over the internet with pay-as-you-go pricing. Users can obtain technology services such as processing power, storage and databases from a cloud provider, eliminating the need for purchasing, operating and maintaining on-premises physical data centers and servers.

A cloud can be private, public or a hybrid. A public cloud sells services to anyone on the internet. A private cloud is a proprietary network or a data center that supplies hosted services to a limited number of people, with certain access and permissions settings. A hybrid cloud offers a mixed computing environment where data and resources can be shared between both public and private clouds. Regardless of the type, the goal of cloud computing is to provide easy, scalable access to computing resources and IT services.

Cloud infrastructure involves the hardware and software components required for the proper deployment of a cloud computing model. Cloud computing can also be thought of as utility computing or on-demand computing.

Cloud computing is the delivery of different services through the Internet. These resources include tools and applications like data storage, servers, databases, networking, and software.

Rather than keeping files on a proprietary hard drive or local storage device, cloud-based storage makes it possible to save them to a remote database. As long as an electronic device has access to the web, it has access to the data and the software programs to run it.

Cloud computing is a popular option for people and businesses for a number of reasons including cost savings, increased productivity, speed and efficiency, performance, and security.

- Cloud computing is the delivery of different services through the Internet, including data storage, servers, databases, networking, and software.

- Cloud storage has grown increasingly popular among individuals who need larger storage space and for businesses seeking an efficient off-site data back-up solution.
- Cloud-based storage makes it possible to save files to a remote database and retrieve them on demand.
- Services can be both public and private—public services are provided online for a fee while private services are hosted on a network to specific clients.
- Cloud security has become an increasingly important field in IT.

Understanding Cloud Computing

Cloud computing is named as such because the information being accessed is found remotely in the cloud or a virtual space. Companies that provide cloud services enable users to store files and applications on remote servers and then access all the data via the Internet. This means the user is not required to be in a specific place to gain access to it, allowing the user to work remotely.

Cloud computing takes all the heavy lifting involved in crunching and processing data away from the device you carry around or sit and work at. It also moves all of that work to huge computer clusters far away in cyberspace. The Internet becomes the cloud, and voilà—your data, work, and applications are available from any device with which you can connect to the Internet, anywhere in the world.

Cloud computing can be both public and private. Public cloud services provide their services over the Internet for a fee. Private cloud services, on the other hand, only provide services to a certain number of people. These services are a system of networks that supply hosted services. There is also a hybrid option, which combines elements of both the public and private services.

Types of Cloud Services

Regardless of the kind of service, cloud computing services provide users with a series of functions including:

- Email
- Storage, backup, and data retrieval
- Creating and testing apps
- Analyzing data
- Audio and video streaming
- Delivering software on demands

Cloud computing is still a fairly new service but is being used by a number of different organizations from big corporations to small businesses, nonprofits to government agencies, and even individual consumers.

Deployment Models

There are various types of clouds, each of which is different from the other. Public clouds provide their services on servers and storage on the Internet. These are operated by third-party companies, who handle and control all the hardware, software, and the general infrastructure. Clients access services through accounts that can be accessed by just about anyone.

Private clouds are reserved for specific clientele, usually one business or organization. The firm's data service center may host the cloud computing service. Many private cloud computing services are provided on a private network.

Hybrid clouds are, as the name implies, a combination of both public and private services. This type of model allows the user more flexibility and helps optimize the user's infrastructure and security.

How does cloud computing work?

Cloud computing lets client devices access rented computing resources, such as data, analytics and cloud applications over the internet. It relies on a network of remote data centers, servers and storage systems that are owned and operated by cloud service providers. The providers are responsible for ensuring the storage capacity, security and computing power needed to maintain the data users send to the cloud.

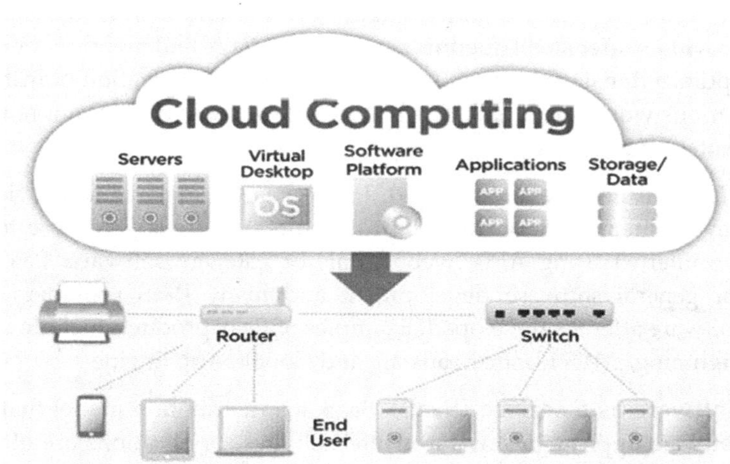

Typically, the following steps are involved in cloud computing:

1. An internet network connection links the front end -- the accessing client device, browser, network and cloud software applications -- with the back end, which consists of databases, servers, operating systems and computers.
2. The back end functions as a repository, storing data accessed by the front end.

3. A central server manages communications between the front and back ends. It relies on protocols to facilitate the exchange of data. The central server uses both software and middleware to manage connectivity between different client devices and cloud servers.
4. Typically, there's a dedicated server for each application or workload.

Cloud computing relies heavily on virtualization and automation technologies. Virtualization lets IT organizations create virtual instances of servers, storage and other resources that let multiple VMs or cloud environments run on a single physical server using software known as a hypervisor. This simplifies the abstraction and provisioning of cloud resources into logical entities, letting users easily request and use these resources. Automation and accompanying orchestration capabilities provide users with a high degree of self-service to provision resources, connect services and deploy workloads without direct intervention from the cloud provider's IT staff.

What are the different types of cloud computing services?

Cloud services can be classified into three general service delivery categories:

1. **Infrastructure as a service (IaaS).** IaaS providers, such as Amazon Web Services (AWS), supply a virtual server instance and storage, as well as application programming interfaces (APIs) that let users migrate workloads to a virtual machine (VM). Users have an allocated storage capacity and can start, stop, access and configure the VM and storage as desired. IaaS providers offer small, medium, large, extra-large and memory- or compute-optimized instances, in addition to enabling customization of instances for various workload needs. The IaaS cloud model is closest to a remote data center for business users.
2. **Platform as a service (PaaS).** In the PaaS model, cloud providers host development tools on their infrastructures. Users access these tools over the internet using APIs, web portals or gateway software. PaaS is used for general software development and many PaaS providers host the software after it's developed. Examples of PaaS products include Salesforce Lightning, AWS Elastic Beanstalk and Google App Engine.
3. **Software as a service (SaaS).** SaaS is a distribution model that delivers software applications over the internet; these applications are often called *web services.* Users can access SaaS applications and services from any location using a computer or mobile device that has internet access. In the SaaS model, users gain access to application software and databases. An example of a SaaS application is Microsoft 365 for productivity and email services.
4. **Function as a service (FaaS).** FaaS, also known as *serverless computing,* lets users run code in the cloud without having to worry about the underlying infrastructure. Users can create and deploy functions that respond to events

or triggers. FaaS abstracts server and infrastructure management, letting developers concentrate solely on code creation.

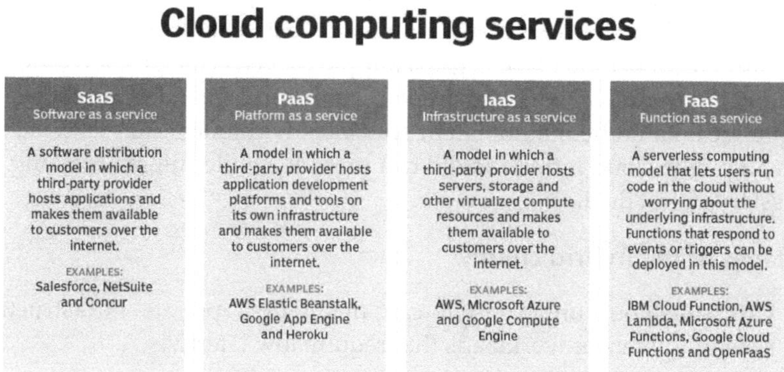

Fig : Main types of Cloud Computing Services

Cloud computing deployment models

There are several cloud computing deployment methods, including the following:

Private cloud

A business's data center delivers private cloud services to internal users. With a private cloud, an organization builds and maintains its own underlying cloud infrastructure. This model offers the versatility and convenience of the cloud, while preserving the management, control and security common to local data centers. Internal users might be billed for services through IT chargeback. Examples of private cloud technologies and vendors include VMware and OpenStack.

Advantages of a private cloud:

- **More flexibility**—your organization can customize its cloud environment to meet specific business needs.
- **More control**—resources are not shared with others, so higher levels of control and privacy are possible.
- **More scalability**—private clouds often offer more scalability compared to on-premises infrastructure.

Public cloud

In the public cloud model, a third-party cloud service provider (CSP) delivers the cloud service over the internet. Public cloud services are sold on demand, typically by the minute or hour, though long-term commitments are available for many services. Customers only pay for the central processing unit cycles, storage or bandwidth they consume. Examples of public CSPs include AWS, Google Cloud Platform (GCP), IBM, Microsoft Azure, Oracle and Tencent Cloud.

Hybrid cloud

A hybrid cloud is a combination of public cloud services and an on-premises private cloud, with orchestration and automation between the two. Companies can run mission-critical workloads or sensitive applications on the private cloud and use the public cloud to handle workload bursts or spikes in demand. The goal of a hybrid cloud is to create a unified, automated, scalable environment that takes advantage of all that a public cloud infrastructure can provide, while still maintaining control over mission-critical data.

Advantages of the hybrid cloud:

- **Control**—your organization can maintain a private infrastructure for sensitive assets or workloads that require low latency.
- **Flexibility**—you can take advantage of additional resources in the public cloud when you need them.
- **Cost-effectiveness**—with the ability to scale to the public cloud, you pay for extra computing power only when needed.
- **Ease**—transitioning to the cloud doesn't have to be overwhelming because you can migrate gradually—phasing in workloads over time.

Multi-cloud

Organizations are increasingly embracing a multi-cloud model, or the use of multiple IaaS providers. This lets applications migrate between different cloud providers or operate concurrently across two or more cloud providers.

Organizations adopt multi-cloud for various reasons, including to help them minimize the risk of a cloud service outage or take advantage of more competitive pricing from a particular provider. It also helps organizations avoid vendor lock-in, letting them switch from one provider to another if needed.

However, multi-cloud deployment and application development can be a challenge because of the differences between cloud providers' services and APIs. Multi-cloud deployments should become easier as cloud providers work toward standardization and convergence of their services and APIs. Industry initiatives such as Open Cloud Computing Interface aim to promote interoperability and simplify multi-cloud deployments.

Benefits of multi-cloud

The benefits of multi-cloud include:

- The ability to use various cloud services for different departments
- The flexibility to easily switch between vendors
- Less downtime with strategically located data centers

Drawbacks of multi-cloud

Migrating to a multi-cloud strategy and managing multiple vendors can be time-intensive and complex. Also, your multi-cloud operations stay siloed, meaning data is not transferred between services. But depending on your needs, this might be a benefit.

Another drawback is the need for your IT team to have the knowledge and skill sets to handle maintenance for multiple cloud deployment methods.

Community cloud

A community cloud, which several organizations share, supports a particular community that has the same concerns, mission, policy, security requirements and compliance considerations. A community cloud is either managed by these organizations or a third-party vendor and can be on or off premises.

Differences between Public and Private Clouds

Public clouds	Private clouds
Multi-tenant infrastructure	Single-tenant infrastructure
No hardware or software purchases required	Operates in your chosen cloud or on-premises data center
Highly scalable and flexible	Customizable and regulation-compliant
Minimal control	Maximum control
Less secure than other options	Highly isolated and secure

Characteristics of cloud computing

Cloud computing has been around for several decades and today's cloud computing infrastructure demonstrates an array of characteristics that have brought meaningful benefits to businesses of all sizes.

Characteristics of cloud computing include the following:

- **Self-service provisioning.** End users can spin up compute resources for almost any type of workload on demand. An end user can provision computing capabilities, such as server time and network storage, eliminating the traditional need for IT administrators to provision and manage compute resources.
- **Elasticity.** Companies can freely scale up as computing needs increase and scale down as demands decrease. This eliminates the need for massive investments in local infrastructure, which might not remain active.
- **Pay per use.** Compute resources are measured at a granular level, letting users pay only for the resources and workloads they use.

- **Workload resilience.** CSPs often deploy redundant resources to ensure resilient storage and to keep users' important workloads running -- often across multiple global regions.
- **Migration flexibility.** Organizations can move certain workloads to or from the cloud or to different cloud platforms automatically.
- **Broad network access.** A user can access cloud data or upload data to the cloud from anywhere with an internet connection using any device.
- **Multi-tenancy and resource pooling.** Multi-tenancy lets several customers share the same physical infrastructures or the same applications, yet still retain privacy and security over their own data. With resource pooling, cloud providers service numerous customers from the same physical resources. The resource pools of the cloud providers should be large and flexible enough so they can service the requirements of multiple customers.
- **Security.** Security is integral in cloud computing and most providers prioritize the application and maintenance of security measures to ensure confidentiality, integrity and availability of data being hosted on their platforms. Along with strong security features, providers also offer various compliance certifications to ensure that their services adhere to industry standards and regulations.

What are the benefits of cloud computing?

Cloud computing provides a variety of benefits for modern business, including the following:

- **Cost management.** Using cloud infrastructure can reduce capital costs, as organizations don't have to spend massive amounts of money buying and maintaining equipment, investing in hardware, facilities or utilities, or building large data centers to accommodate their growing businesses. In addition, companies don't need large IT teams to handle cloud data center operations because they can rely on the expertise of their cloud providers' teams. Cloud computing also cuts costs related to downtime. Since downtime rarely happens in cloud computing, companies don't have to spend time and money to fix issues that might be related to downtime.
- **Data and workload mobility.** Storing information in the cloud means users can access it from anywhere with any device with just an internet connection. That means users don't have to carry around USB drives, an external hard drive or multiple CDs to access their data. They can access corporate data via smartphones and other mobile devices, letting remote employees stay current with co-workers and customers. End users can easily process, store, retrieve and recover resources in the cloud. In addition, cloud vendors provide all the upgrades and updates automatically, saving time and effort.

- **Business continuity and disaster recovery (BCDR).** All organizations worry about data loss. Storing data in the cloud guarantees that users can always access their data even if their devices, such as laptops or smartphones, are inoperable. With cloud-based services, organizations can quickly recover their data in the event of natural disasters or power outages. This benefits BCDR and helps ensure that workloads and data are available even if the business suffers damage or disruption.
- **Speed and agility.** Cloud computing facilitates rapid deployment of applications and services, letting developers swiftly provision resources and test new ideas. This eliminates the need for time-consuming hardware procurement processes, thereby accelerating time to market.
- **Environmental sustainability.** By maximizing resource utilization, cloud computing can help to promote environmental sustainability. Cloud providers can save energy costs and reduce their carbon footprint by consolidating workloads onto shared infrastructure. These providers often operate large-scale data centers designed for energy efficiency.

What are the disadvantages of cloud computing?

Despite the clear upsides to relying on cloud services, cloud computing has its own challenges for IT professionals:

- **Cloud security.** Security is often considered the greatest challenge organizations face with cloud computing. When relying on the cloud, organizations risk data breaches, hacking of APIs and interfaces, compromised credentials and authentication issues. Furthermore, there's a lack of transparency regarding how and where sensitive information entrusted to the cloud provider is handled. Security demands careful attention to cloud configurations and business policy and practice.
- **Unpredictable costs.** Pay-as-you-go subscription plans for cloud use, along with scaling resources to accommodate fluctuating workload demands, can make it difficult to define and predict final costs. Cloud costs are also frequently interdependent, with one cloud service often using one or more other cloud services -- all of which appear in the recurring monthly bill. This can create additional unplanned cloud costs.
- **Lack of expertise.** With cloud-supporting technologies rapidly advancing, organizations are struggling to keep up with the growing demand for tools and employees with the proper skills and knowledge needed to architect, deploy and manage workloads and data in a cloud.
- **IT governance difficulties.** The emphasis on do-it-yourself in cloud computing can make IT governance difficult, as there's no control over provisioning, deprovisioning and management of infrastructure operations. This can make it challenging for organizations to properly manage risks and security, IT compliance and data quality.

- **Compliance with industry laws.** When transferring data from on-premises local storage into cloud storage, it can be difficult to manage compliance with industry regulations through a third party. It's important to know where data and workloads are actually hosted to maintain regulatory compliance and proper business governance.
- **Management of multiple clouds.** Every cloud is different, so multi-cloud deployments can disjoint efforts to address more general cloud computing challenges.
- **Cloud performance.** Performance -- such as latency -- is largely beyond the control of the organization contracting cloud services with a provider. Network and provider outages can interfere with productivity and disrupt business processes if organizations aren't prepared with contingency plans.
- **Cloud migration.** The process of moving applications and other data to the cloud often causes complications. Migration projects frequently take longer than anticipated and go over budget. The issue of workload and data repatriation -- moving from the cloud back to a local data center -- is often overlooked until unforeseen costs or performance problems arise.
- **Vendor lock-in.** Often, switching between cloud providers can cause significant issues. This includes technical incompatibilities, legal and regulatory limitations and substantial costs incurred from sizable data migrations.

Cloud computing examples

Cloud computing has evolved and diversified into a wide array of offerings and capabilities designed to suit almost any conceivable business need. Examples of cloud computing capabilities and diversity include the following:

- **Google Docs, Microsoft 365.** Users can access Google Docs and Microsoft 365 via the internet. Users can be more productive because they can access work presentations and spreadsheets stored in the cloud anytime from anywhere on any device.
- **Email, Calendar, Skype, WhatsApp.** Emails, calendars, Skype and WhatsApp take advantage of the cloud's ability to provide users with access to data remotely so they can examine their data on any device, whenever and wherever they want.
- **Zoom.** Zoom is a cloud-based software platform for video and audio conferencing that records meetings and saves them to the cloud, letting users access them anywhere and at any time. Another common communication and collaboration platform is Microsoft Teams.
- **AWS Lambda.** Lambda lets developers run code for applications or back-end services without having to provision or manage servers. The pay-as-you-go model constantly scales with an organization to accommodate

real-time changes in data usage and data storage. Other examples of major cloud providers that also support serverless computing capabilities include Google Cloud Functions and Microsoft Azure Functions.

- **Salesforce.** Salesforce is a cloud-centric customer relationship management platform designed to assist businesses in overseeing their sales, marketing and customer service operations.

Cloud computing use cases

How is the cloud actually used? The myriad services and capabilities found in modern public clouds have been applied across countless use cases, such as the following:

- **Testing and development.** Ready-made, tailored environments can expedite timelines and milestones.
- **Production workload hosting.** Organizations are using the public cloud to host live production workloads. This requires careful design and architecture of cloud resources and services needed to create an adequate operational environment for the workload and its required level of resilience.
- **Big data analytics.** Remote data centers through cloud storage are flexible and scalable and can provide valuable data-driven insights. Major cloud providers offer services tailored to big data analytics and projects, such as Amazon EMR and Google Cloud Dataproc.
- **IaaS.** IaaS lets companies host IT infrastructures and access compute, storage and network capabilities in a scalable manner. Pay-as-you-go subscription models are cost-effective, as they can help companies save on upfront IT costs.
- **PaaS.** PaaS can help companies develop, run and manage applications more easily and flexibly, at a lower cost than maintaining a platform on premises. PaaS services can also increase the development speed for applications and enable higher-level programming.
- **Hybrid cloud.** Organizations have the option to use the appropriate cloud -- private or public -- for different workloads and applications to optimize cost and efficiency according to the circumstance.
- **Multi-cloud.** Using multiple different cloud services from separate cloud providers can help subscribers find the best cloud service fit for diverse workloads with specific requirements.
- **Storage.** Large amounts of data can be stored remotely and accessed easily. Clients only have to pay for storage that they actually use.
- **Disaster recovery.** Cloud offers faster recovery than traditional on-premises . Furthermore, it's offered at lower costs.

- **Data backup.** Cloud backup options are generally easier to use. Users don't have to worry about availability and computing capacity, and the cloud provider manages data security.
- **Artificial intelligence as a service.** Cloud computing lets individuals without formal knowledge or expertise in data sciences reap the benefits of AIaaS. For example, a web developer might create a facial recognition app with their web development skills. AI is available as a service in the cloud and accessible via the API. This lets users automate routine tasks, saving time and personnel costs. Businesses can also enhance decision-making by using AI to predict outcomes based on historical datasets.
- **Internet of things.** Cloud computing simplifies the processing and management of data from IoT devices. Cloud platforms offer the scalability and processing capacity required to handle the enormous amounts of data produced by IoT devices, facilitating real-time analytics and decision-making. For example, an IoT device system such as Google Nest or Amazon Alexa can collect data on how much energy is used inside a smart home. The device can then use cloud computing to analyze the gathered data and make recommendations to the homeowner on how to reduce energy consumption.

Cloud computing vs. traditional web hosting

Given the many different services and capabilities of the public cloud, there has been some confusion between cloud computing and major uses, such as web hosting. While the public cloud is often used for web hosting, the two are quite different. Significant innovations in virtualization and distributed computing, as well as improved access to high-speed internet, have accelerated interest in cloud computing.

The distinct characteristics of cloud computing that differentiate it from traditional web hosting include the following:

1. With cloud computing, users can access large amounts of computing power on demand. It's typically sold by the minute or the hour. With traditional hosting, users typically pay for a set amount of storage and processing power. Since resources are limited, businesses can look into virtual private servers or dedicated hosting as their business and demands grow.
2. Cloud computing is elastic, meaning users can have as much or as little of a service as they want at any given time. However, with traditional hosting, scalability is often constrained, particularly in shared hosting. Shared hosting involves multiple websites sharing resources on a single server, potentially causing performance issues and slower website speeds if one site encounters a sudden surge in traffic.
3. Service is fully managed by the provider on cloud computing platforms; the consumer needs nothing but a PC and internet access. While shared

traditional hosting is also fully managed by the provider, typically users are required to control their website from a user-friendly interface such as cPanel.

4. In comparison to traditional hosting, cloud hosting is more reliable. Cloud providers maintain redundant infrastructure and operate across numerous data centers, reducing downtime and increasing availability. Traditional hosting, on the other hand, is based on a single server, making it more prone to hardware failures and higher downtime threats.

5. Both cloud hosting and traditional hosting entail security considerations. Cloud hosting providers invest significantly in security measures to safeguard data and infrastructure. However, certain organizations might find traditional hosting more suitable, as it provides greater control over security measures and can accommodate specific security requirements.

Cloud computing service providers

The cloud service market has no shortage of providers. The three largest public CSPs -- AWS, GCP and Microsoft Azure -- have established themselves as dominant players in the industry. According to the Synergy Research Group, at the end of 2022, these three vendors made up 66% of the worldwide cloud infrastructure market.

Other major CSPs include the following:

- Alibaba.
- Citrix.
- IBM.
- Oracle.
- Rackspace.
- Salesforce.
- SAP.
- VMware.

When selecting a cloud service vendor, organizations should consider certain things. First, the actual suite of services can vary between providers, and business users must select a provider offering services -- such as big data analytics or AI services -- that support the intended use case.

Though cloud services typically rely on a pay-per-use model, different providers often have variations in their pricing plans to consider. Furthermore, if the cloud provider will be storing sensitive data, an organization should also consider the physical location of the provider's servers.

Naturally, reliability and security should be top priorities. A provider's service-level agreement should specify a level of service uptime that's satisfactory to client business needs. When considering different cloud vendors, organizations should pay close attention to what technologies and configuration settings are used to secure sensitive information.

What is a cloud service provider?

A cloud service provider, or CSP, is an IT company that provides on-demand, scalable computing resources like computing power, data storage, or applications over the internet. Typically, cloud-based service models are defined as IaaS (infrastructure as a service), PaaS (platform as a service), or SaaS (software as a service).

Types of cloud providers

CSPs offer a variety of services, but typically they fall under three types of cloud service providers:

IaaS providers

IaaS solutions provide access to IT infrastructure components that you would normally have in your data center, eliminating the need to procure, configure, or manage your own. This includes resources like compute, networks, storage, data management, virtualization, and operating systems. While the IaaS model increases flexibility, lowers costs, and speeds up times to market, customers are still responsible for managing and maintaining their own applications and responding to issues.

PaaS providers

PaaS solutions build off IaaS, providing tools and services to create and deploy applications. PaaS incorporates operating systems, middleware, and runtime environments into the application stack and also manages any hardware or other assets related to the underlying infrastructure.

SaaS providers

SaaS solutions are perhaps the most familiar and make up the fastest growing services in the cloud market. CSPs deliver ready-to-use applications and are responsible for maintaining and managing everything, from hardware and maintenance to development, scaling, and delivery. For example, you probably use many of the SaaS productivity applications in Google Workspace every day, such as Gmail, Calendar, Docs, and Drive.

While these are the most common categories, they do not fully describe all of the types of services currently offered by CSPs. Many top cloud service providers are continuously developing new cloud-based services around emerging technologies like containerization, edge computing, machine learning, and Kubernetes.

Types of cloud service providers can also be categorized based on whether they deliver services via public cloud infrastructure or use private, hybrid, or multicloud deployment models.

Cloud service provider examples

The CSP market includes cloud providers of all shapes and sizes. The big three—Google Cloud, Microsoft Azure, and Amazon Web Services (AWS)—are considered the established leaders. However, there are a host of other smaller or niche players that offer cloud services as well, including IBM, Alibaba, Oracle, Red Hat, DigitalOcean, and Rackspace.

Increasingly, companies are choosing to mix and match cloud services from different CSPs for different requirements rather than choosing one provider for everything. In most cases, choosing to work with multiple cloud providers helps businesses select the best cloud capabilities for their unique use cases. For instance, different providers may specialize in certain areas like data analytics and AI services or offer better support for legacy environments or broader compute options.

CSPs that follow an open approach to the cloud give their customers the freedom to combine the services and providers that best fit their needs, allowing them to move workloads to and from on-premises and between cloud providers at any time.

Benefits of using a cloud service provider

Business agility

CSPs shoulder many of the responsibilities of maintaining, repairing, and securing hardware and software. This gives IT teams more time to focus on development and significantly speeds up time to market.

Reduced costs

Cloud service providers offer pay-as-you-go access to cloud services. You only pay for what you consume, helping to reduce the necessary upfront investment needed to build and maintain IT infrastructure.

Scalability

Public cloud providers offer nearly infinite resources, which can be scaled up or down quickly to accommodate demand, such as an unexpected surge in traffic or adding capacity to support new business growth.

Reliability

Cloud service providers offer expertise and experience, ensuring reliable cloud infrastructure. They must also guarantee a certain level of uptime and performance to fulfill their service-level agreements (SLAs).

Centralized location

A CSP can also help you centralize and integrate your services and software in a single place, making it easier for users to access, analyze, and activate valuable data and resources.

Future-proof systems

CSPs are constantly evolving to make services available around emerging technologies. The right cloud provider can help you ensure your systems and tools keep up with the latest advancements.

Improved mobility

The services and resources of cloud service providers are accessible from any physical location with a working network connection. Remote workers can securely access applications and files from anywhere as long as they have an internet connection.

Disaster recovery

CSPs follow robust redundancy and resilience plans for business continuity to help minimize downtime and enable faster recovery of services even after severe disruptions.

Challenges of using a cloud service provider

While there are many advantages to working with a cloud provider, there are some considerations to keep in mind.

Here are some of the key things for organizations to consider:

- **Complex contracts:** You will need to negotiate contracts and SLAs with each cloud provider you use. Multiple providers and vendors can lead to complex SLA relationships with different parameters and guaranteed service expectations.
- **Vendor lock-in:** Some cloud providers do not integrate well with competitor products and services. Becoming too dependent on a single provider can make it difficult to migrate data and workloads to another technology stack without incurring high costs, incompatibilities, or even legal constraints.
- **Security responsibility:** CSPs follow a shared responsibility model, which means cloud security is implemented by both the cloud provider and the customer. Poor understanding of a provider's responsibilities and your own can lead to substantial security risks or security breaches.

Along with these specific concerns when working with cloud service providers, your organization will also need to have a solid cloud migration strategy in place and be ready to tackle new complexity when it comes to monitoring and managing cloud environments.

How to choose a cloud service provider

Cloud adoption is rarely a straightforward, one-size-fits-all journey. Choosing the right provider or providers, will depend on your unique needs and technology requirements.

Below is a list of some of the key factors to consider when evaluating cloud service providers:

1. Cost

While it shouldn't be the only reason for choosing a cloud provider, cost is often one of the primary factors in making a decision. It's helpful to think about both the direct costs of service usage (upfront and pay-as-go) and any indirect costs, such as hiring talent or retiring on-premises systems.

2. Digital capabilities and processes

Beyond the cloud products and services available, you should assess how well a CSP can help you meet your current and future IT needs. It's helpful to consider how easy it is for you to manage and deploy services and what integration is available for existing business-critical applications. Other important considerations include whether they use standard interfaces and APIs, event and change management, and support for hybrid and multicloud models.

3. Trust

Trust in your cloud service provider is a prerequisite for a successful relationship. You should make an honest assessment of what you really need from a CSP and consider whether a provider can meet those expectations. For instance, does the CSP have a good market reputation? What level of cloud experience and technical knowledge do they have? Are they financially stable? And will they be able to provide the support and guidance you need to reach your goals?

4. Open ecosystem

Increasingly, proprietary solutions do not suit the technical requirements of modern business. You should evaluate cloud service providers on how "open" they are. For example, look at whether you have options to build, migrate, and deploy your applications across multiple environments, both in the cloud or on-premises. A top cloud service provider should leverage open source technologies and interoperable solutions that ensure consistency and effective management wherever your workloads may be.

5. Security

Finally, you should consider the security practices of a cloud provider. Your CSP must demonstrate that they can keep your business and customer data safe. This includes evaluating everything from security infrastructure to security policies and identity management to data backup and retention. It's also essential to find out what controls exist to ensure the physical security of their data center, such as environmental safeguards, disaster recovery, and documented business continuity plans.

Cloud computing security

Security remains a primary concern for businesses contemplating cloud adoption -- especially public cloud adoption. Public CSPs share their underlying hardware infrastructure between numerous customers, as the public cloud is a multi-tenant environment. This environment demands significant isolation between logical compute resources. At the same time, access to public cloud storage and compute resources is guarded by account login credentials.

Many organizations bound by complex regulatory obligations and governance standards are still hesitant to place data or workloads in the public cloud for fear of outages, loss or theft. However, this resistance is fading, as logical isolation has proven reliable and the addition of data encryption and various identity and access management tools have improved security within the public cloud.

Ultimately, the responsibility for establishing and maintaining a secure cloud environment falls to the individual business user who is responsible for building the workload's architecture -- the combination of cloud resources and services in which the workload runs -- and using the security features that the cloud provider offers.

History of cloud computing

The history and evolution of cloud computing date back to the 1950s and 1960s.

- **1950s.** Companies started using large mainframe computers, but due to the expense, not every organization could afford to purchase one. So, during the late 1950s and early 1960s, a process called *time sharing* was developed to make more efficient use of expensive processor time on the central mainframe. Time sharing lets users access numerous instances of computing mainframes simultaneously, maximizing processing power and minimizing downtime. This idea represents the first use of shared computing resources, the foundation of modern cloud computing.
- **1960s.** The origins of delivering computing resources using a global network are, for the most part, rooted in 1969 when American computer scientist J.C.R. Licklider helped create the Advanced Research Projects Agency Network, the so-called precursor to the internet. Licklider's goal was to connect computers across the globe in a way that would let users access programs and information from any location.
- **1970s.** Cloud computing began to take a more tangible shape with the introduction of the first VMs, letting users run more than one computing system within a single physical setup. The functionality of these VMs led to the concept of virtualization, which had a major influence on the progress of cloud computing.
- **1980s.** In the 1970s and 1980s, Microsoft, Apple and IBM developed technologies that enhanced the cloud environment and advanced the use of cloud server and server hosting.
- **1990s.** In 1999, Salesforce became the first company to deliver business applications from a website.
- **2000s.** In 2006, Amazon launched AWS, providing services like computing and storage in the cloud. Following suit, the other major tech players,

including Microsoft and Google, launched their own cloud offerings to compete with AWS.

- **2010s.** Microsoft launched Azure in 2010 and Office 365 in 2011. Also, the Docker container technology was first released in this decade. Microservices and serverless platforms were also introduced in this timeframe when the Google App Engine was launched in 2008, followed by AWS Lambda in 2015.
- **2020 and beyond.** The growth of serverless computing is anticipated to continue and edge computing will become more and more significant in the upcoming years. Edge computing devices are expected to improve and become more interconnected as AI and machine learning technologies advance.

Future of cloud computing and emerging technologies

Cloud computing is expected to see substantial breakthroughs and the adoption of new technologies. Back in its "2020 Data Attack Surface Report," Arcserve predicted that there will be 200 zettabytes of data stored in the cloud by 2025.

Some major trends and key points that are shaping the future of cloud computing include the following:

- Organizations are increasingly migrating mission-critical workloads to public clouds. One reason for this shift is that business executives who want to ensure that their companies can compete in the new world of digital transformation are demanding the public cloud.
- Business leaders are also looking to the public cloud to take advantage of its elasticity, modernize internal computer systems, and empower critical business units and their DevOps teams. Cloud providers, such as IBM and VMware, are concentrating on meeting the needs of enterprise IT, in part by removing the barriers to public cloud adoption that caused IT decision-makers to shy away from fully embracing the public cloud previously.
- Generally, when contemplating cloud adoption, many enterprises have mainly focused on new cloud-native applications -- that is, designing and building applications specifically intended to use cloud services. They haven't been willing to move their most mission-critical apps into the public cloud. However, these organizations are beginning to realize that the cloud is ready for the enterprise if they select the right cloud platforms.
- Cloud providers are locked in ongoing competition for cloud market share, so the public cloud continues to evolve, expand and diversify its range of services. This has resulted in public IaaS providers offering more than common compute and storage instances. For example, serverless, or event-driven, computing is a cloud service that executes specific functions, such as image processing and database updates. Traditional cloud deployments require users to establish a compute instance and load code into that instance.

Then, the user decides how long to run -- and pay for -- that instance. With serverless computing, developers simply create code and the cloud provider loads and executes that code in response to real-world events so users don't have to worry about the server or instance aspect of the cloud deployment. Users only pay for the number of transactions that the function executes. AWS Lambda, Google Cloud Functions and Azure Functions are examples of serverless computing services.

- Public cloud computing also lends itself well to big data processing, which demands enormous compute resources for relatively short durations. Cloud providers have responded with big data services, including Google BigQuery for large-scale data warehousing and Microsoft Azure Data Lake Analytics for processing huge data sets.
- Another crop of emerging cloud technologies and services relates to AI and machine learning. These technologies provide a range of cloud-based, ready-to-use AI and machine learning services for client needs. Examples of these services include Amazon Machine Learning, Amazon Lex, Amazon Polly, Google Cloud Machine Learning Engine and Google Cloud Speech API.
- The connection between blockchain and cloud computing is poised to strengthen, as companies recognize blockchain's potential to improve operational efficiency, security and transparency. This trend is further supported by increased investment and the expansion of blockchain-as-a-service platforms.

Cloud Computing :

Cloud Computing referred as the accessing and storing of data and provide services related to computing over the internet. It simply referred as it remote services on the internet manage and access data online rather than any local drives. The data can be anything like images, videos, audios, documents, files etc.

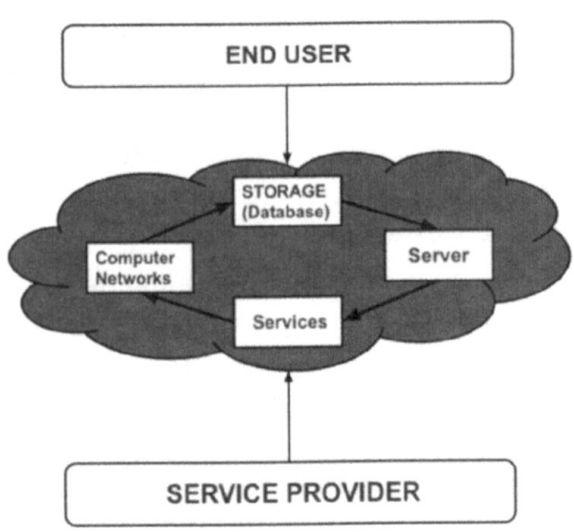

Cloud Computing Service Provider's :

Cloud computing is in huge demand so, big organization providing the service like_Amazon AWS, Microsoft Azure, Google Cloud, Alibaba cloud etc. are some Cloud Computing service Provider.

Historyof Cloud Computing :

In this, we will discuss the history of Cloud computing. And also cover the history of client server computing, distributed computing, and cloud computing.

- Before Computing was come into existence, client Server Architecture was used where all the data and control of client resides in Server side. If a single user want to access some data, firstly user need to connect to the server and after that user will get appropriate access. But it has many disadvantages. So, After Client Server computing, Distributed Computing was come into existence, in this type of computing all computers are networked together with the help of this, user can share their resources when needed. It also has certain limitations. So in order to remove limitations faced in distributed system, cloud computing was emerged.

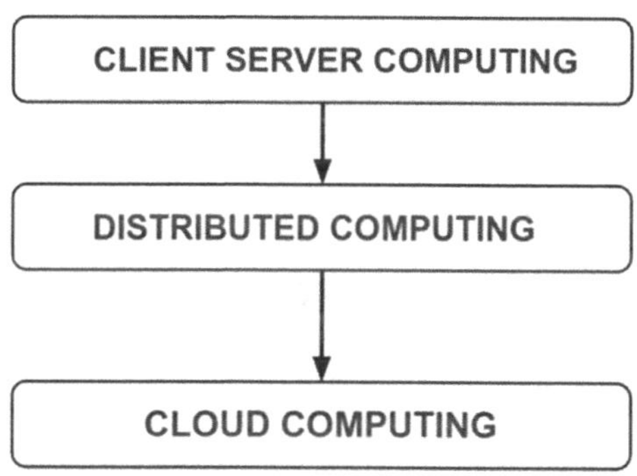

- During 1961, John MacCharty delivered his speech at MIT that "Computing Can be sold as a Utility, like Water and Electricity." According to John MacCharty it was a brilliant idea. But people at that time don't want to adopt this technology. They thought the technology they are using efficient enough for them. So, this concept of computing was not appreciated much so and very less will research on it. But as the time fleet the technology caught the idea after few years this idea is implemented. So, this is implemented by Salesforce.com in 1999.
- This company started delivering an enterprise application over the internet and this way the boom of Cloud Computing was started.
- In 2002, Amazon started Amazon Web Services (AWS), Amazon will provide storage, computation over the internet. In 2006 Amazon will launch Elastic Compute Cloud Commercial Service which is open for Everybody to use.

- After that in 2009, Google Play also started providing Cloud Computing Enterprise Application as other companies will see the emergence of cloud Computing they also started providing their cloud services. Thus, in 2009, Microsoft launch Microsoft Azure and after that other companies like Alibaba, IBM, Oracle, HP also introduces their Cloud Services. In today the Cloud Computing become very popular and important skill.

Evolution of Cloud Computing

Cloud computing is all about renting computing services. This idea first came in the 1950s. In making cloud computing what it is today, five technologies played a vital role. These are distributed systems and its peripherals, virtualization, web 2.0, service orientation, and utility computing.

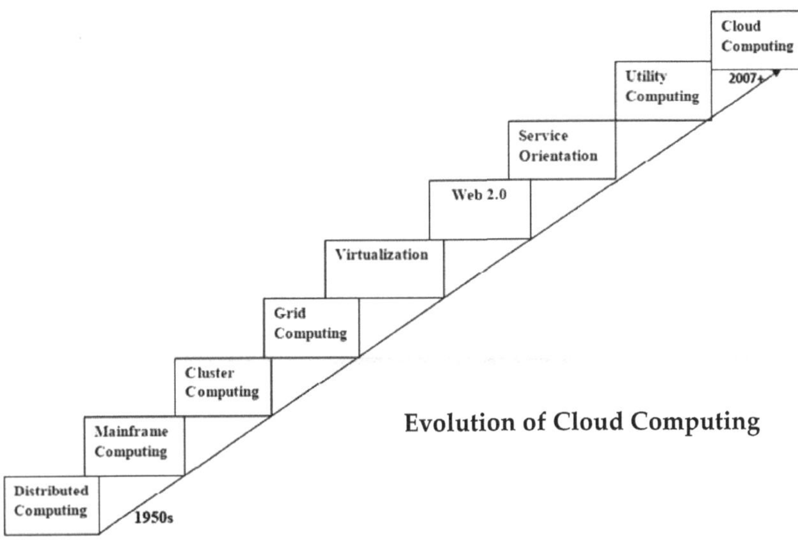

Evolution of Cloud Computing

- **Distributed Systems:**

  It is a composition of multiple independent systems but all of them are depicted as a single entity to the users. The purpose of distributed systems is to share resources and also use them effectively and efficiently. Distributed systems possess characteristics such as scalability, concurrency, continuous availability, heterogeneity, and independence in failures. But the main problem with this system was that all the systems were required to be present at the same geographical location. Thus to solve this problem, distributed computing led to three more types of computing and they were-Mainframe computing, cluster computing, and grid computing.

- **Mainframe computing:**

  Mainframes which first came into existence in 1951 are highly powerful and reliable computing machines. These are responsible for handling large data such as massive input-output operations. Even today these are used for bulk processing tasks such as online transactions etc. These systems have almost no downtime with high fault tolerance. After distributed computing, these increased the processing capabilities of the system. But these were very expensive. To reduce this cost, cluster computing came as an alternative to mainframe technology.

- **Cluster computing:**

  In 1980s, cluster computing came as an alternative to mainframe computing. Each machine in the cluster was connected to each other by a network with high bandwidth. These were way cheaper than those mainframe systems. These were equally capable of high computations. Also, new nodes could easily be added to the cluster if it was required. Thus, the problem of the cost was solved to some extent but the problem related to geographical restrictions still pertained. To solve this, the concept of grid computing was introduced.

- **Grid computing:**

  In 1990s, the concept of grid computing was introduced. It means that different systems were placed at entirely different geographical locations and these all were connected via the internet. These systems belonged to different organizations and thus the grid consisted of heterogeneous nodes. Although it solved some problems but new problems emerged as the distance between the nodes increased. The main problem which was encountered was the low availability of high bandwidth connectivity and with it other network associated issues. Thus. cloud computing is often referred to as "Successor of grid computing".

- **Virtualization:**

  It was introduced nearly 40 years back. It refers to the process of creating a virtual layer over the hardware which allows the user to run multiple instances simultaneously on the hardware. It is a key technology used in cloud computing. It is the base on which major cloud computing services such as Amazon EC2, VMware vCloud, etc work on. Hardware virtualization is still one of the most common types of virtualization.

- **Web 2.0:**

  It is the interface through which the cloud computing services interact with the clients. It is because of Web 2.0 that we have interactive and dynamic web pages. It also increases flexibility among web pages. Popular examples of web 2.0 include Google Maps, Facebook, Twitter, etc. Needless to say, social media is possible because of this technology only. It gained major popularity in 2004.

- **Service orientation:**

  It acts as a reference model for cloud computing. It supports low-cost, flexible, and evolvable applications. Two important concepts were introduced in this computing model. These were Quality of Service (QoS) which also includes the SLA (Service Level Agreement) and Software as a Service (SaaS).

- **Utility computing:**

  It is a computing model that defines service provisioning techniques for services such as compute services along with other major services such as storage, infrastructure, etc which are provisioned on a pay-per-use basis.

Cloud Computing, which is one of the demanding technology of the current time and which is giving a new shape to every organization by providing on demand virtualized services/resources. Starting from small to medium and medium to large, every organization use cloud computing services for storing information and accessing it from anywhere and any time only with the help of internet. In this article, we will know more about the internal architecture of cloud computing. Transparency, scalability, security and intelligent monitoring are some of the most important constraints which every cloud infrastructure should experience. Current research on other important constraints is helping cloud computing system to come up with new features and strategies with a great capability of providing more advanced cloud solutions.

Cloud Computing Architecture :

The cloud architecture is divided into 2 parts i.e.

1. Frontend
2. Backend

The below figure represents an internal architectural view of cloud computing.

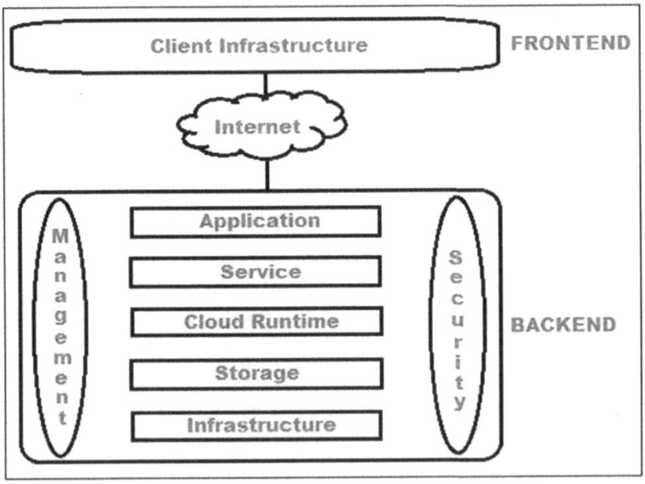

Architecture of cloud computing is the combination of both SOA (Service Oriented Architecture) and EDA (Event Driven Architecture). Client infrastructure, application, service, runtime cloud, storage, infrastructure, management and security all these are the components of cloud computing architecture.

1. Frontend :

Frontend of the cloud architecture refers to the client side of cloud computing system. Means it contains all the user interfaces and applications which are used by the client to access the cloud computing services/resources. For example, use of a web browser to access the cloud platform.

- **Client Infrastructure** – Client Infrastructure is a part of the frontend component. It contains the applications and user interfaces which are required to access the cloud platform.
- In other words, it provides a GUI(Graphical User Interface) to interact with the cloud.

2. Backend :

Backend refers to the cloud itself which is used by the service provider. It contains the resources as well as manages the resources and provides security mechanisms. Along with this, it includes huge storage, virtual applications, virtual machines, traffic control mechanisms, deployment models, etc.

1. **Application –**

   Application in backend refers to a software or platform to which client accesses. Means it provides the service in backend as per the client requirement.

2. **Service –**

   Service in backend refers to the major three types of cloud based services like SaaS, PaaS and IaaS. Also manages which type of service the user accesses.

3. **Runtime Cloud-**

   Runtime cloud in backend provides the execution and Runtime platform/ environment to the Virtual machine.

4. **Storage –**

   Storage in backend provides flexible and scalable storage service and management of stored data.

5. **Infrastructure –**

   Cloud Infrastructure in backend refers to the hardware and software components of cloud like it includes servers, storage, network devices, virtualization software etc.

6. **Management –**

   Management in backend refers to management of backend components like application, service, runtime cloud, storage, infrastructure, and other security mechanisms etc.

7. **Security –**

   Security in backend refers to implementation of different security mechanisms in the backend for secure cloud resources, systems, files, and infrastructure to end-users.

8. Internet –

Internet connection acts as the medium or a bridge between frontend and backend and establishes the interaction and communication between frontend and backend.

9. Database–

Database in backend refers to provide database for storing structured data, such as SQL and NOSQL databases. Example of Databases services include Amazon RDS, Microsoft Azure SQL database and Google CLoud SQL.

10. Networking–

Networking in backend services that provide networking infrastructure for application in the cloud, such as load balancing, DNS and virtual private networks.

11. Analytics–

Analytics in backend service that provides analytics capabillities for data in the cloud, such as warehousing, bussness intellegence and machine learning.

CLOUD STORAGE

Cloud Storage is a mode of computer data storage in which digital data is stored on servers in off-site locations. The servers are maintained by a third-party provider who is responsible for hosting, managing, and securing data stored on its infrastructure. The provider ensures that data on its servers is always accessible via public or private internet connections.

Cloud Storage enables organizations to store, access, and maintain data so that they do not need to own and operate their own data centers, moving expenses from a capital expenditure model to operational. Cloud Storage is scalable, allowing organizations to expand or reduce their data footprint depending on need.

How does Cloud Storage work?

Cloud Storage uses remote servers to save data, such as files, business data, videos, or images. Users upload data to servers via an internet connection, where it is saved on a virtual machine on a physical server. To maintain availability and provide redundancy, cloud providers will often spread data to multiple virtual machines in data centers located across the world. If storage needs increase, the cloud provider will spin up more virtual machines to handle the load. Users can access data in Cloud Storage through an internet connection and software such as web portal, browser, or mobile app via an application programming interface (API).

Cloud Storage is available in four different models:

Public

Public Cloud Storage is a model where an organization stores data in a service provider's data centers that are also utilized by other companies. Data in public

Cloud Storage is spread across multiple regions and is often offered on a subscription or pay-as-you-go basis. Public Cloud Storage is considered to be "elastic" which means that the data stored can be scaled up or down depending on the needs of the organization. Public cloud providers typically make data available from any device such as a smartphone or web portal.

Private

Private Cloud Storage is a model where an organization utilizes its own servers and data centers to store data within their own network. Alternatively, organizations can deal with cloud service providers to provide dedicated servers and private connections that are not shared by any other organization. Private clouds are typically utilized by organizations that require more control over their data and have stringent compliance and security requirements.

Hybrid

A hybrid cloud model is a mix of private and public cloud storage models. A hybrid cloud storage model allows organizations to decide which data it wants to store in which cloud. Sensitive data and data that must meet strict compliance requirements may be stored in a private cloud while less sensitive data is stored in the public cloud. A hybrid cloud storage model typically has a layer of orchestration to integrate between the two clouds. A hybrid cloud offers flexibility and allows organizations to still scale up with the public cloud if need arises.

Multicloud

A multicloud storage model is when an organization sets up more than one cloud model from more than one cloud service provider (public or private). Organizations might choose a multicloud model if one cloud vendor offers certain proprietary apps, an organization requires data to be stored in a specific country, various teams are trained on different clouds, or the organization needs to serve different requirements that are not stated in the servicers' Service Level Agreements. A multicloud model offers organizations flexibility and redundancy.

Why is cloud computing important?

For developers, cloud computing provides increased amounts of storage and processing power to run the applications they develop. Cloud computing also enables new ways to access information, process and analyze data, and connect people and resources from any location anywhere in the world. In essence, it takes the lid off the box; with cloud computing, developers are no longer boxed in by physical constraints.

For IT departments, cloud computing offers more flexibility in computing power, often at lower costs. With cloud computing, IT departments don't have to engineer for peak-load capacity, because the peak load can be spread out among the external assets in the cloud. And, because additional cloud resources are always at the ready, companies no longer have to purchase assets (servers, workstations, and the like) for infrequent intensive computing tasks. If you need more processing power, it's always there in the cloud—and accessible on a cost-efficient basis.

For end users, cloud computing offers all these benefits and more. An individual using a web-based application isn't physically bound to a single PC, location, or network. His applications and documents can be accessed wherever he is, whenever he wants. Gone is the fear of losing data if a computer crashes. Documents hosted in the cloud always exist, no matter what happens to the user's machine.

And then there's the benefit of group collaboration, for both individuals and organizations. Users from around the world can collaborate on the same documents, applications, and projects, in real time. It's a whole new world of collaborative computing, all enabled by the notion of cloud computing.

Advantages and Disadvantages of Cloud Computing

Advantages of Cloud Computing

As we all know that Cloud computing is trending technology. Almost every company switched their services on the cloud to rise the company growth.

Here, we are going to discuss some important advantages of Cloud Computing-

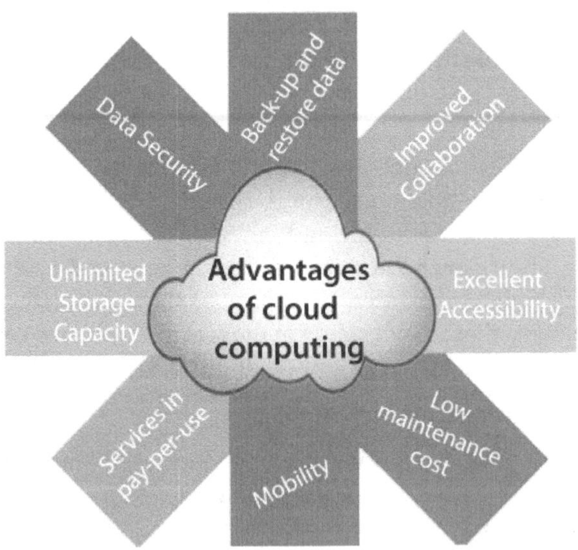

1) Back-up and restore data

Once the data is stored in the cloud, it is easier to get back-up and restore that data using the cloud.

2) Improved collaboration

Cloud applications improve collaboration by allowing groups of people to quickly and easily share information in the cloud via shared storage.

3) Excellent accessibility

Cloud allows us to quickly and easily access store information anywhere, anytime in the whole world, using an internet connection. An internet cloud infrastructure increases organization productivity and efficiency by ensuring that our data is always accessible.

4) Low maintenance cost

Cloud computing reduces both hardware and software maintenance costs for organizations.

5) Mobility

Cloud computing allows us to easily access all cloud data via mobile.

6) IServices in the pay-per-use model

Cloud computing offers Application Programming Interfaces (APIs) to the users for access services on the cloud and pays the charges as per the usage of service.

7) Unlimited storage capacity

Cloud offers us a huge amount of storing capacity for storing our important data such as documents, images, audio, video, etc. in one place.

8) Data security

Data security is one of the biggest advantages of cloud computing. Cloud offers many advanced features related to security and ensures that data is securely stored and handled.

Disadvantages of Cloud Computing

A list of the disadvantage of cloud computing is given below -

1) Internet Connectivity

As you know, in cloud computing, every data (image, audio, video, etc.) is stored on the cloud, and we access these data through the cloud by using the internet connection. If you do not have good internet connectivity, you cannot access these data. However, we have no any other way to access data from the cloud.

2) Vendor lock-in

Vendor lock-in is the biggest disadvantage of cloud computing. Organizations may face problems when transferring their services from one vendor to another. As different vendors provide different platforms, that can cause difficulty moving from one cloud to another.

3) Limited Control

As we know, cloud infrastructure is completely owned, managed, and monitored by the service provider, so the cloud users have less control over the function and execution of services within a cloud infrastructure.

4) Security

Although cloud service providers implement the best security standards to store important information. But, before adopting cloud technology, you should be aware that you will be sending all your organization's sensitive information to a third party, i.e., a cloud computing service provider. While sending the data on the cloud, there may be a chance that your organization's information is hacked by Hackers.

CLOUD SERVICE PROVIDERS:

Top 10 cloud service providers:

1. Amazon Web Services (AWS)
2. Microsoft Azure
3. Google Cloud
4. Alibaba Cloud
5. IBM Cloud
6. Oracle
7. Salesforce
8. SAP
9. Rackspace Cloud
10. VMWare

What is a Cloud–based application?

- Cloud apps, also known as cloud-based or SaaS (Software as a Service) apps, are applications that are hosted and run on remote servers (cloud infrastructure).
- Users access cloud apps through web browsers, similar to web apps, but the entire application resides and operates in the cloud.
- Cloud apps are typically designed to be scalable, flexible, and available on demand.
- These apps can store and process large amounts of data in cloud databases and benefit from cloud computing resources for efficient performance.
- Cloud apps often offer subscription-based pricing models and are maintained, updated, and secured by the service provider.
- Examples of cloud apps include Google Workspace (formerly G Suite), Microsoft Office 365, and Salesforce.

Types of Cloud-Based Applications:

Cloud-based applications are software applications that are hosted on remote servers and accessed through the Internet. They offer several advantages over traditional on-premises applications, including scalability, flexibility, and affordability.

There are three main types of cloud-based applications:

Software as a Service (SaaS):

SaaS applications are the most common type of cloud-based application. They are fully hosted and managed by the cloud provider, so users can access them through a web browser or mobile app without having to install any software on their own devices. Popular SaaS applications include Google Workspace, Microsoft Office 365, and Salesforce.

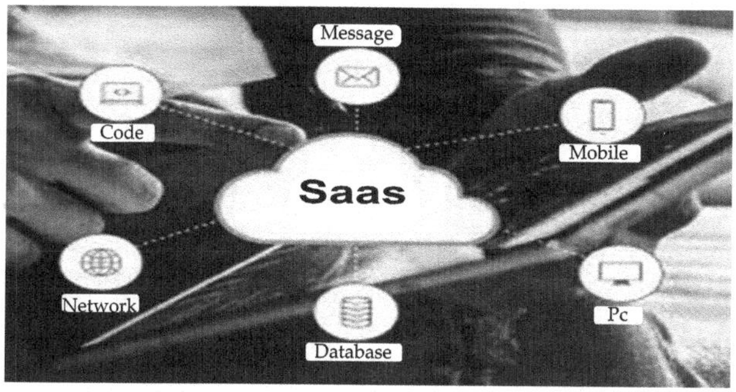

Platform as a Service (PaaS):

PaaS applications provide a platform for developers to build, deploy, and manage their own applications. They include the tools and infrastructure needed for application development, such as programming languages, databases, and middleware. Popular PaaS applications include Heroku and Azure App Service.

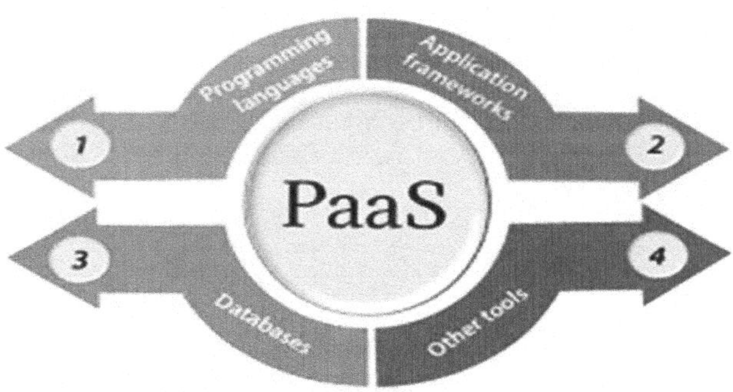

Infrastructure as a Service (IaaS):

IaaS applications provide the basic computing resources needed to run applications, such as virtual machines, storage, and networking. Users can use these resources to build and deploy their own applications or to run existing applications on the cloud. Popular IaaS applications include Amazon Web Services (AWS), Microsoft Azure, and Google Cloud Platform (GCP).

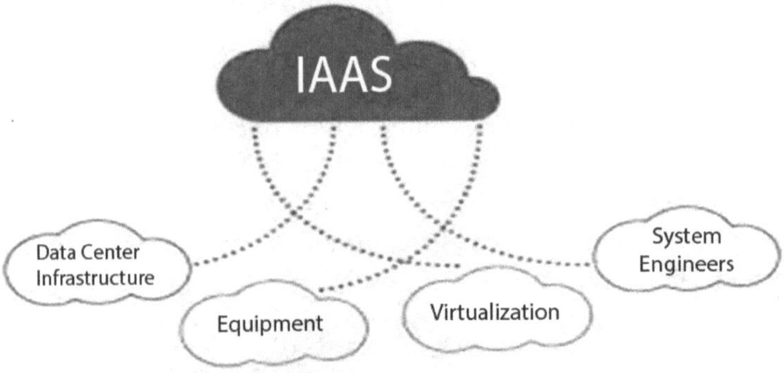

In addition to these three main types, there are also a number of other cloud-based applications available, such as:

Database as a Service (DBaaS):

DBaaS applications provide managed database services, such as MySQL, PostgreSQL, and Oracle. They allow users to create, manage, and scale their databases without having to worry about the underlying infrastructure.

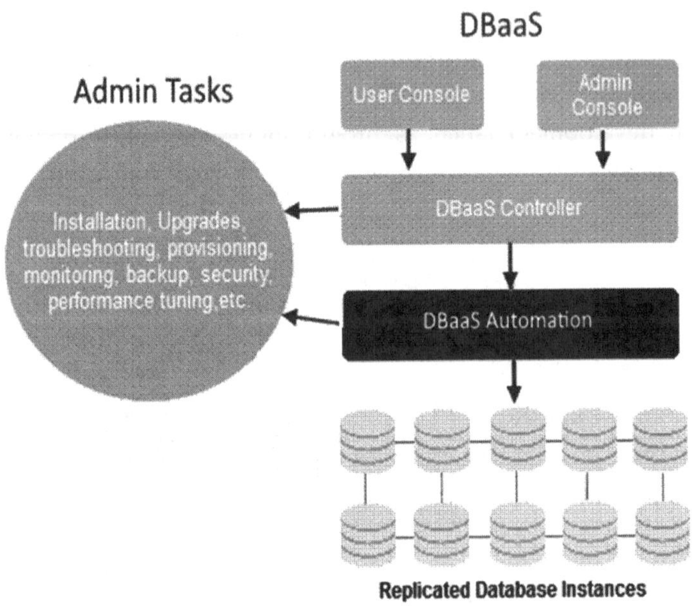

Storage as a Service (SaaS):

SaaS applications provide cloud-based storage for data, such as files, images, and videos. They can be used to back up data, share data with others, or host web content. Popular SaaS applications include Amazon S3, Google Cloud Storage, and Dropbox.

Content Delivery Network (CDN):

CDN applications distribute web content across a network of servers around the world. This improves the performance and reliability of web applications by serving content from the server that is closest to the user. Popular CDN applications include Cloudflare and Akamai.

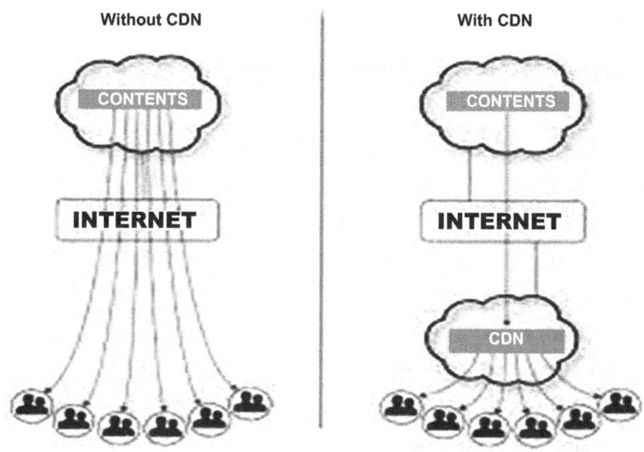

Benefits of Cloud-based Application

The cloud-based application is helpful for any business since remote servers take care of most data storage and processing issues.

1. **Scalability**: Cloud apps can easily scale their resources up or down based on demand, ensuring smooth performance even during peak usage periods. This scalability allows businesses to accommodate a growing number of users without investing in additional infrastructure.

2. **Cost-Effective**: Cloud apps follow a subscription-based or pay-as-you-go model, eliminating the need for upfront hardware and software investments. Businesses can avoid the costs associated with maintaining on-premises servers and infrastructure, leading to cost savings.

   Accessibility: Cloud apps are accessible from any device with an internet connection, allowing users to work or access data from virtually anywhere. This enhances productivity and facilitates remote collaboration, especially in today's global and mobile workforce.

3. **Automatic Updates and Maintenance**: Cloud app providers handle regular updates, maintenance, and security patches, ensuring that users always have access to the latest features and bug fixes. This eliminates the burden of software maintenance on users and IT departments.

4. **Data Security**: Reputable cloud service providers employ advanced security measures, such as encryption, multi-factor authentication, and regular data backups, to protect user data. This level of security can often exceed what individual businesses can provide on their own.

What is a Web-based Application?

- Web apps are software applications that run on web browsers, making them accessible from any device with an internet connection.
- These apps are typically hosted on web servers and are delivered to users through the Internet.
- Users interact with web apps by accessing their URLs in a web browser, without needing to download or install any software.
- Web apps are written in programming languages such as HTML, CSS, and JavaScript, and they rely on web technologies like AJAX for dynamic content.
- Examples of web apps include online banking portals, social media platforms, and email clients.
- Web apps do not require local installations or frequent updates, as changes are made on the server side.

Types of Web-based Applications

The vast IT business market for web-based software is already shifting to web applications. As we all know, many multinational companies use web apps but don't use similar types of apps. Every business's needs vary, so the need for customized apps differs. Thus, we are here to present six different types of web applications.

Single page application

Single-page applications are dynamic web applications that do not require browser reloads and function as a single unit of a website application. It is applied to all the business and technological strategies in the client-side browser, so the web apps perform pretty fast.

Multi-Page application

It includes multiple pages and reloads the whole page whenever the user navigates to a different page, known as a multiple-page app. The companies like Google Docs and Trello provide multiple services and products. Thus, it allows users to connect in various ways.

Static web application

This web app is not adaptable and is built with HTML and CSS. But changing content on static web apps is tricky because you need to download HTML code, change it and upload it to the server. The webmaster does this task.

Animated web application

With this application's usage, you can display your material with animation effects. These applications provide creativity and different designs unavailable in other applications.

Progressive web application

These web technologies are created using various languages like HTML, CSS, and JavaScript. However, it is also called cross-platform web applications that use the latest browser APIs, progressive enhancement methods, and features. It offers an experience like a native mobile app.

Content management system (CMS)

In these web apps, the content is constantly updated. A CMS administrator is the person who grants permission for upgrades, modifications, and updates. WordPress is a popular CMS platform for creating commercial and personal blogs, news, etc.

Advantages of using Web-based Applications

1. **Cross-Platform Compatibility**: Web apps are compatible with multiple operating systems and devices, making them accessible to a broader audience. Users can access web apps on Windows, macOS, Linux, Android, iOS, and other platforms without the need for platform-specific development.
2. **No Installation Required**: Web apps do not require installation or downloads since they run within web browsers. This ease of access saves users from the hassle of installing and updating software on their devices.
3. **Faster Deployment**: Web apps can be deployed quickly, as they do not need to go through the app store approval process. Once developed, web apps can be published and made available to users instantly.
4. **Real-Time Updates**: Changes made to web apps are reflected in real-time for all users, ensuring that everyone is always using the latest version of the application. This allows businesses to implement updates and improvements rapidly.
5. **Easy Maintenance and Bug Fixing**: Since web apps operate centrally on web servers, developers can quickly address issues and bugs without requiring users to download or install updates. This streamlined maintenance process benefits both developers and end-users.

Overall, both Cloud apps and Web apps offer numerous advantages, catering to different use cases and business needs. The choice between the two depends on factors such as scalability requirements, cost considerations, accessibility preferences, and the need for centralized maintenance and updates.

Web Application

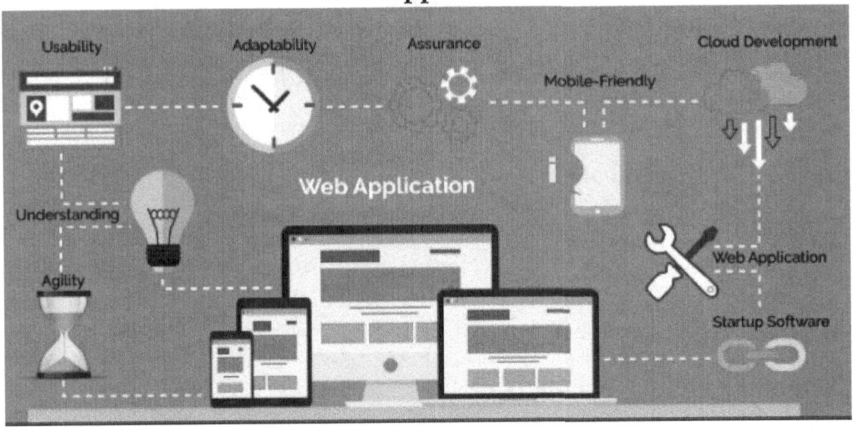

Web Application Architecture Components

Typically, a web application architecture diagram comprises three core components:

1) **Web Browser**: The browser, client-side, or front-end component is the key component that interacts with the user, receives the input, and manages the presentation logic while controlling user interactions with the application. User inputs are validated as well if required.

2) **Web Server:** The web server, also known as the backend or server-side component, handles the business logic and processes the user requests by routing the requests to the right component and managing the entire application operations. It can run and oversee requests from a wide variety of clients.

3) **Database Server**: The database server provides the required data for the application. It handles data-related tasks. In a multi-tiered architecture, database servers can manage business logic with the help of stored procedures.

What is a 3-Tier Architecture?

In a traditional 2-Tier architecture, there are two components: the client side system or the user interface and a backend system, which is usually a database server. The business logic is incorporated into the user interface or the database server. The downside of 2-tier architecture is that the performance decreases with an increased number of users. Moreover, the direct interaction of the database and the user device raises security concerns. Railway reservation systems and content management systems are a couple of applications that are usually built using this architecture.

There are three layers of a 3-Tier architecture:

1. Presentation layer / Client Layer
2. Application Layer / Business Layer
3. Data Layer

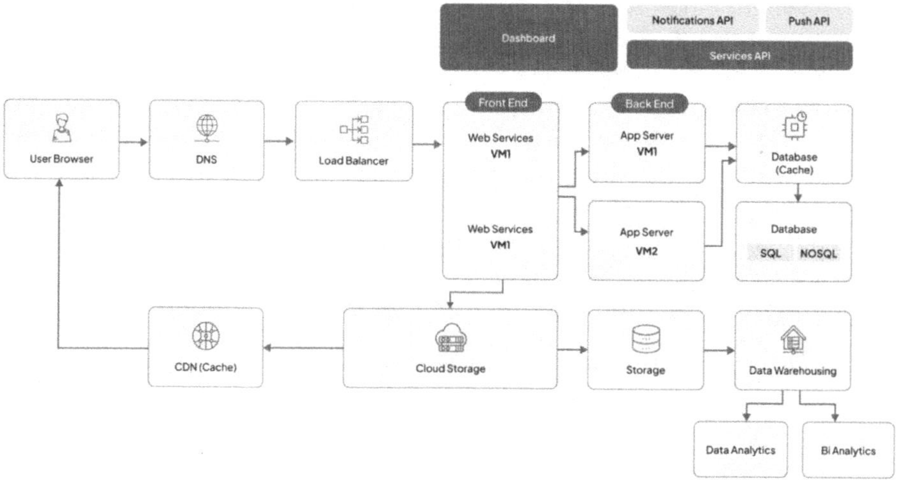

Application Layer: Web Server

What is a web server? Simply put, a web server runs one or more websites or web apps. The web server uses HyperText Transfer Protocol (HTTP) and other protocols to view user requests via a browser. It processes them by applying business logic and delivering the requested content to the end-user.

A Web Server can be a hardware device or a software program.

- **Hardware Web Server**: A computer device connected to the internet and contains web server software and web app components such as images, HTML docs, JS files and CSS style sheets.
- **Software Web Server**: It is software that understands URLs and HTTP protocols. Users can access it via domain names to receive the requested content.

While a static web server delivers the content to the browser as it is, a dynamic web server updates data before delivering it.

Apache is a popular open-source web server from the Apache Software Foundation. It was developed by Robert McCool in C and XML in 1995. Apache is based on a process-driven model wherein every request results in the creation of a new thread. The modular design of Apache allows you to scale individual resources easily. With minimal configuration, you can manage even large traffic. It works on MacOS, Windows, and Linux environments. However, Linux is the most preferred environment for Apache.

While it uses a file system to process static content, dynamic content is processed within the server. Using .htaccess files, you can perform additional configurations to the server settings. Security is good. It offers support via IRC, Stack Overflow, and mailing lists.

NGINX is another popular web server called 'Engine X.' Developed by Igor Sysoev in 2004, NGINX quickly became popular. It operates on an event-driven model wherein thousands of requests are processed within a single thread, delivering more with minimal resources. It uses PHP to serve static resources and serves static content 2. 5 times faster than Apache. Dynamic content is served via external processes. Regarding interpreting requests, Apache passes the file system location while NGINX passes the URI. This feature extends the NGINX capability as a load balancer, HTTP Cache, and proxy server.

While it supports Unix-based OS, Windows compatibility is limited. You can't make additional configurations. The smaller codebase offers better security. Dynamic modules are not supported. Along with mailing lists and IRC, Forum is available, too.

NGINX has an edge over Apache as it serves as a web server and a proxy server. The event-driven approach that processes thousands of requests in a single thread offers greater performance, speed and cost-effectiveness.

Presentation Layer: Client-side Component (Front-end)

The client-side component of a web application architecture diagram enables users to interact with the server and the backend service via a browser. The code resides in the browser, receives requests, and presents the user with the required information. This is where UI/UX design, dashboards, notifications, configurational settings, layout, and interactive elements come into the picture.

HTML

HTML or Hypertext Markup Language is a popular standard markup language that enables developers to structure web page contents using a series of page elements. Developed by Tim Berners-lee and released in 1993, HTML quickly evolved and became the standard markup language globally.

CSS

CSS or Cascading Style Sheets is a popular style sheet language that lets developers separate website content and layout for sites developed using markup languages. Using CSS, you can define a style for elements and reuse them multiple times. Similarly, you can apply one style across multiple sites. It is simple and easy to learn. You can apply a style for a single element, an entire webpage, or the entire website. It is device-friendly, too.

Browser compatibility and security are two areas that raise a concern. Similarly, different versions of CSS also create confusion. Developers are advised to check the compatibility before making any changes to the design.

JavaScript

JavaScript or JS is the most popular client-side programming language, which has been used by more than 90% of websites recently. It was designed by Brendan Eich of Netscape in 1995. JavaScript uses a simple, easy-to-learn syntax. The language is so popular that every browser uses a JS engine to run JavaScript code on devices.

It is easy to insert JS code on any web page, making it highly interoperable. It allows you to create rich interfaces to deliver a better UI/UX experience. Being on the client side, JS reduces the server load as well.

However, developers should be careful about the security as the code is executed on the client side, which hackers can sometimes exploit.

React

React is an open-source JavaScript that has gained popularity in recent years. It was developed by Jordan Walke of Facebook in 2013. React benefits enable developers to easily create high-quality dynamic web applications with minimal code and effort.

ReactJS is easy to learn and use. There is extensive documentation and plenty of handy tools available for developers. The code is reusable. ReactJS uses a virtual DOM, which means concerned elements are updated when a change is made instead of the entire DOM tree being rewritten. It improves efficiency and optimizes memory usage. ReactJS uses a one-way Data flow, meaning changes made to the "child" elements do not affect the "parent" element. The code is easy to test and SEO-friendly.

On the downside, the ReactJS development environment is highly dynamic, so developers should proactively monitor the changes and quickly adapt new skills to leverage React. Moreover, React technologies are improving rapidly. However, the documentation cannot catch up with this pace. A critical area of concern is that ReactJS focuses on the UI part, and you need to depend on other libraries for client-side functionalities.

Vue.js

Vue.js is an open-source JavaScript framework written by Evan You in 2014. This framework enables developers to build web, desktop, and mobile UI interfaces easily. Vue.js comes with handy tools that serve basic programming needs. The tool is lightweight for downloads and installation. It is easy to learn and use. It uses a virtual DOM so that when a user interacts with an element, the browser does not have to render the entire page, only the element. The speed and performance are good. It uses a two-way data binding model, allowing you to track data and update related components more efficiently. The components are reusable. It easily integrates with existing apps. The documentation is concise and well-structured. Community support is good.

While the community support is good, Vue.js is largely adopted by Chinese companies such as Alibaba and Xiaomi. As such, forums and discussions are often done in Chinese, creating a language barrier for English-speaking people. The tool kit suits basic projects but offers limited support for large-scale projects. Its flexibility can pose quality issues on large projects as well. GitLab, Alibaba, and Adobe are popular companies that use Vue.js.

Angular.js

Angular is an open-source web app framework developed by Google in 2016. It is a complete rewrite of the Angular.js framework. As of now, it is one of the most popular front-end development frameworks available in the market.

NGModules is the building block of Angular, offering all functionalities for developing applications such as components, modules, templates, directives, service and dependency injections, routing, etc. It helps developers quickly build prototypes. It uses plain HTML templates. Testing is quick and easy owing to the dependency injection architecture style.

Angular uses two-way data binding, which makes DOM manipulations easy and quick. The CPU-performance enhancing features make it a good choice for large-scale apps. It comes with various plugin tools out of the box. Coming from the IT giant Google, Angular enjoys vibrant community support. The popularity and market value imply that you'll find highly skilled Angular professionals in the market.

However, the hierarchical structure can sometimes make debugging a challenge. Concepts such as Inversion Of Control, Dependency Lookup, and Dependency Injection require a steep learning curve. You need JavaScript installed on the machine to run Angular. While Two-way data binding is a great feature, it can slow performance on older and legacy devices. Integrating legacy infrastructure with Angular is a concern, too.

If you want to pick the best ones out of these front-end development tools, React and Vue.js are highly recommended. React is a lightweight tool with the best developer functions, enabling you to build quality software quickly.

Vue.js is a view-oriented product that is lightweight, easy to use and comes with a powerful set of developer tools. You just need to load the interface and add JavaScript to get started.

Application Layer: Server-side Component (Back-end)

The server-side component is the key component of the web application architecture diagram that receives user requests, performs business logic, and delivers the required data to the front-end systems. It contains servers, databases, web services, etc.

Node.js

Node.js is an open-source, cross-platform runtime environment developed by Ryan Dahl. It was built on Google Chrome V8 Engine to run network and server-side applications and was released in 2009. Developers use JavaScript to build node.js applications and run them on node.js runtime using Windows, macOS and Linux platforms.

Node.js is highly popular as it offers a rich library of JavaScript modules, enabling developers to build quality applications quickly. Node.js doesn't buffer data and executes code super fast. It is event-driven, asynchronous, and highly scalable on a single thread. Node.js best suits apps that are data streaming, data-intensive, I/O bound, and JSON-API based. Paypal, Uber, eBay, and GoDaddy are some popular applications that Node.js powers. It doesn't suit apps that are CPU-intensive.

Java

Java is one of the most popular and effective programming languages ever. Written by James Gosling in 1995, Java is an object-oriented and class-based programming language that enables developers to write code and run it on any platform using a Java virtual machine (JVM) environment. It means you don't need Java on the target machine. The language is easy to learn, code, compile, and debug. Being platform-agnostic, Java programs are cost-effective to build and run. Leveraging OOPS concepts, you can reuse objects in other programs. It doesn't work with explicit pointers, so unauthorized memory access is avoided. It supports multi-threading, portability, automatic garbage collection, distributed networking, etc.

On the downside, Java requires significant memory space. Owing to the JVM abstraction, the programs run slower as well. There is no backup, either. The UX/UI is not attractive. However, the benefits of developing server-side components of the web application architecture diagram in Java outweigh the drawbacks.

Python

Python is an open-source high-level programming language written by Guido Van Rossum and released in 1991. Today, it is a fast-growing and popular programming language and a strong alternative for building web app architecture in Java. It is simple to learn and develop and rich in features. This dynamically typed language is highly flexible and suitable for small and large web application projects and a variety of segments such as mobile apps, video games, AI programming, etc. It allows you to do more with less code, so you can quickly build and test prototypes that increase productivity. Python offers an extensive library that contains code for almost every type of program. With its vibrant community support always available.

However, compared with modern languages, Python is slower. Threading issues are another concern. Python is not native to mobile apps. Google, Spotify, Instagram, and Facebook are popular IT giants that use Python.

PHP Laravel

PHP Laravel is a PHP framework that helps developers easily build custom web apps. It is an open-source framework that is quite popular among other PHP frameworks. PHP Laravel is an MVC (Model, View and Controller)-based framework.

Laravel syntax is elegant and expressive. With extensive built-in functions and structures, developers can easily write code and deploy apps faster. It improves performance and speed. Documentation is good. An important advantage of PHP Laravel is its automated testing feature that helps you to test and debug errors in the initial stage. Automatic tasks and scheduling are available, too. Object Relational Mapping support is elegant. PHP Laravel offers cross-site request forgery tokens that offer security. It is scalable and cost-effective.

However, product upgrades can create problems at times. The Laravel Composer can be improved. Community support is not so great since it's a relatively new product. PHP Laravel best suits medium and small organizations.

Go

Go Programming language comes from the IT giant Google, which gives it considerable strength in the first place. It was written by Robert Griesemer, Ken Thompson, and Rob Pike in 2009. Go is also known as Golang. It is similar to the C language and is easy to learn and build. With no virtual runtimes, Go code compiles faster and produces smaller binary. It uses static typing and interface types. The standard library offers a range of built-in functions along with testing support. Garbage collection is available. Concurrent programming is easy compared to other languages.

On the downside, the library support is not adequate. The community support is not receptive on some occasions. Implicit interfaces can be challenging to manage. Without generics, the reuse of code is not easy.

NET

NET is a software development framework developed by Microsoft and released in 2001 for desktop and web applications. Coming from the IT giant Microsoft,. NET quickly became popular for developing a variety of software products.

NET comes in 3 flavors:

- **NET Framework**: It is the first product that was specific for Windows platform
- **NET Core**: The .NET Core was released in 2016 as a cross-platform solution to accommodate macOS and Linux platforms.
- **Xamarin:** Xamarin was not developed by Microsoft but was acquired by the company in 2016. Xamarin extends the .NET platform to support native mobile application development.

.NET Standard is a single base class library for .NET Framework, .NET Core, and Xamarin implementations.

NET is an Object Oriented Programming (OOP) model and uses a modular structure that enables developers to break down code into smaller pieces and seamlessly build and manage software products using CI/CD pipelines. It offers a robust yet simple caching system that improves speed and performance. Automatic monitoring in ASP.NET is an added advantage. The Visual Studio IDE is a single IDE that helps developers monitor and manage entire development operations from a single pane. Being language-agnostic and platform-agnostic, it allows you to use a variety of development environments. Deployment and management of code is flexible and easy. .NET comes with extensive documentation and community support.

However, object-relational support is limited. Memory leaks are a concern that needs careful attention while coding apps. Being heavily dependent on Microsoft results in vendor lock-in and higher licensing costs. .NET best suits enterprise products that require high scalability and cross-platform solutions.

Ruby

Ruby is a popular programming language developed by Yukihiro Matsumoto of Japan in 1995. Time efficiency is one of Ruby's biggest advantages. Combined with the Rails framework, it allows developers to build and deploy apps quickly. The tool offers an extensive library and helpful tools. It comes with built-in security to mitigate risks related to SQL injections, cross-site scripting software (XSS), and cross-site request forged (CSRF). Ruby has a supportive community and good documentation.

While developers can quickly build applications, the speed of the applications is a concern. That said, this problem mainly affects large-scale apps. Small and medium-sized organizations don't have any issues in this field.

Ruby is not as popular as Java or Python. For this reason, it is not easy to find quality professionals for this segment. Airbnb, GitHub, Bloomberg, and Etsy are popular companies that use Ruby.

Among the server-side development tools, Node.js and Python are highly recommended. Node.js is an easy-to-learn, lightweight, developer-friendly, highly scalable, and extensible cross-platform solution. Python uses simple syntax, focuses on natural language, and makes it easy to write and execute code faster. The community is also very supportive.

Application Layer: Application Programming Interface (API)

Application Programming Interface (API) is not a technology- it's a concept that enables developers to access specific data and functions of software. Simply put, it is a mediator that allows apps to communicate with each other. It comprises protocols, tools, and subroutine definitions required to build apps.

For instance, when you sign in to an application, the app calls the API to retrieve your account details and credentials. The application will contact the corresponding servers to receive this information and return this data to the user app. A web API is an API available over the web via HTTP protocol. It can be built using technologies such as .NET and Java.

With APIs, developers don't have to create everything from scratch but use existing functions exposed as an API to increase productivity and gain faster time to market. By reducing development efforts, APIs significantly reduce development costs. It also improves collaboration and connectivity across the ecosystem while enhancing customer experience.

There are different types of APIs

- **RESTful API**: Representational State Transfer API in lightweight JSON format. It is highly scalable, dependable and delivers fast performance, making it the most popular API.
- **SOAP**: Simple Object Access Protocol uses XML for data transmission. It requires more bandwidth and advanced security
- **XML-RPC**: Extensible Markup Language – Remote Procedure Calls use specific XML format for data transmission
- **JSON-RPC**: It uses JSON format for data transmission

Application Layer: Server Instance / Cloud Instance

Servers or cloud instances are important to a web application architecture diagram. A cloud instance is a virtual server instance built, delivered, and hosted using a public or a private cloud and is accessible over the Internet. It works as a physical server that seamlessly moves across multiple devices or deploys multiple instances on a single server. As such, it is highly dynamic, scalable and cost-effective. You can automatically replace servers without application downtime. You can easily deploy and manage web applications in any environment using cloud instances.

Data Layer: Database

A database is a key component of a web application that stores and manages information for a web app. Using a function, you can search, filter and sort information based on user request and present the required info to the end user. They allow role-based access to maintain data integrity.

When choosing a database for your architecture of a web app, the size, speed, scalability, and structure are the four aspects that require your consideration. For structured data, SQL-based databases are a good choice. it suits financial apps wherein data integrity is a key requirement.

To handle unstructured data, NoSQL is a good option. It suits apps wherein the nature of incoming data is not predictable. Key Value databases associate each value with a key and suit databases that store user profiles, reviews, blog comments, etc. For analytics, Wide Column databases are a good choice

A key difference between a cloud application and a web application:

The key difference between cloud applications and web applications lies in their architecture and deployment methods. While both types of applications are accessible over the Internet, they serve different purposes and utilize distinct technologies.

Architecture:

Cloud Application: A cloud application, also known as Software-as-a-Service (SaaS), is a type of software that operates on remote cloud servers. Users access the application through a web browser, just like web applications. However, unlike traditional web applications, the data processing and logic for cloud applications occur on the cloud servers, and the application can be accessed from anywhere with an internet connection.

Web Application: A web application is a software program that runs on a web server and is accessed by users through a web browser. The application's logic and data processing occurs on the web server, and the user interface is rendered on the client-side (web browser). These applications are typically built using programming languages like HTML, CSS, and JavaScript.

Data Processing Location:

Cloud Application: Imagine cloud applications as the tech wizards working their magic on remote cloud servers. When you use them, the heavy lifting happens at these distant data centers.

Web Application: Web applications, on the other hand, run like neighborhood businesses, with data processing taking place on specific web servers, and you accessing them through your web browser.

Scalability:

Cloud Application: Cloud applications are like elastic bands. They can easily stretch and accommodate more users by adding extra server resources whenever there's a surge in demand.

Web Application: Web applications might resemble traditional shops that can only serve a limited number of customers at a time. Their scalability depends on the capacity of the single web server they run on.

Accessibility:

Cloud Application: Cloud applications are your globetrotters; they can be accessed from anywhere with an internet connection, making them incredibly flexible, especially for remote and mobile users.

Web Application: Web applications, however, have a specific address – the web server they call home. They're accessible through web browsers, but their availability is tied to where that server is.

Data Security:

Cloud Application: Think of cloud applications as well-guarded fortresses. Cloud service providers invest heavily in security, offering robust measures to protect your data.

Web Application: Web applications are like houses you have to secure yourself. Their security depends on the organization hosting them, which means you might need to put in extra effort to safeguard your data.

Deployment and Maintenance:

Cloud Application: Deploying and maintaining cloud applications is like hiring a reliable service. The responsibility largely falls on the cloud service provider, which simplifies your life.

Web Application: Web applications are more like tending to your own garden. You'll need to manage the web server, which can involve more effort for upkeep and updates.

Cost Structure:

Cloud Application: Cloud applications often come with flexible payment plans. You can subscribe or pay as you go, offering cost predictability and adaptability.

Web Application: Web applications may require more upfront investment, as organizations need to handle their infrastructure and maintenance.

Customization:

Cloud Application: Cloud applications can sometimes be like buying a pre-designed home. You get what's available, with limited customization options.

Web Application: Web applications are the architects' dream. You can shape and mold them to fit your unique needs and preferences

UNIT II
CLOUD ARCHITECTURE AND MODEL

Distributed Computing System Models

Distributed computing refers to a system where processing and data storage is distributed across multiple devices or systems, rather than being handled by a single central device.

Architecture Styles in Distributed Systems

Distributed systems are a set of autonomous computers that appears to be a single coherent system to its users from outside. There are actually two different types of systems that exist in prospective of computers:

Centralized System :

- A centralized system consists of a single machine.
- All calculations are done by a particular computer.
- Its performance is low as the workload is not divided.
- There is also no machine present in backup if the original computer system fails.

Distributed Systems :

- A distributed system consists of multiple machines.
- All computation work is divided among the different systems.
- Its performance is high as the workload is divided among different computers to efficiently use their capacity.
- There are systems present in backup, so if the main system fails then work will not stop.

A distributed system contains multiple nodes that are physically separate but mixed together using the communication networks.

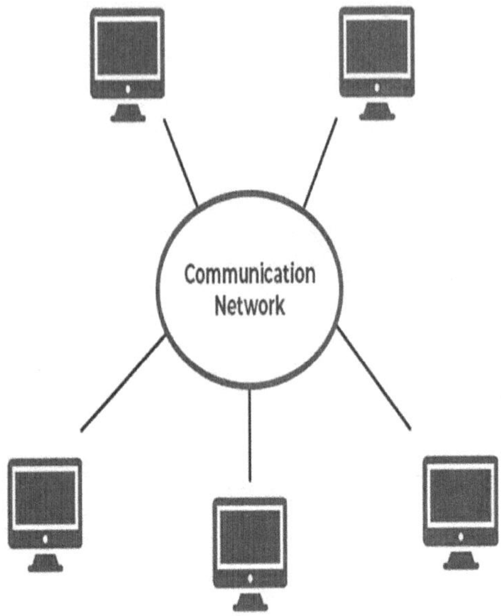

Architecture Styles:

To show different arrangement styles among computers Architecture styles are proposed.

1. Layered Architecture:

In Layered architecture, different components are organised in layers. Each layer communicates with its adjacent layer by sending requests and getting responses. The layered architecture separates components into units. It is an efficient way of communication. Any layer can not directly communicate with another layer. A layer can only communicate with its neighbouring layer and then the next layer transfers information to another layer and so on the process goes on.

In some cases, layered architecture is in cross-layer coordination. In a cross-layer, any adjacent layer can be skipped until it fulfils the request and provides better performance results. Request flow from top to bottom(downwards) and response flow from bottom to top(upwards). The advantage of layered architecture is that each layer can be modified independently without affecting the whole system. This type of architecture is used in Open System Interconnection (OSI) model.

To the layers on top, the layers at the bottom offer a service. While the response is transmitted from bottom to top, the request is sent from top to bottom. This method has the advantage that calls always follow a predetermined path and that each layer is simple to replace or modify without affecting the architecture as a whole.

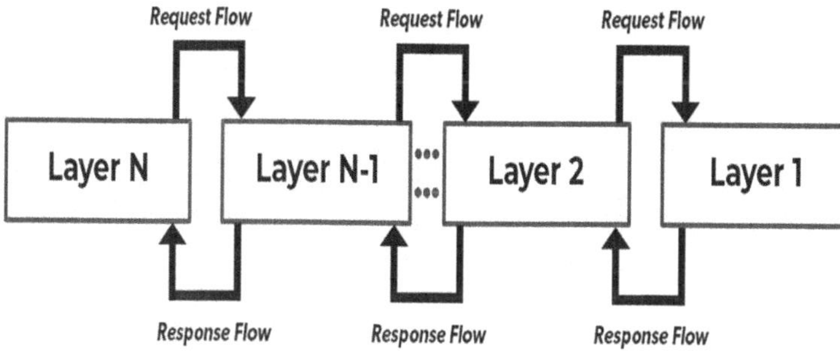

2. Object-Oriented Architecture:

In this type of architecture, components are treated as objects which convey information to each other. Object-Oriented Architecture contains an arrangement of loosely coupled objects. Objects can interact with each other through method calls. Objects are connected to each other through the Remote Procedure Call (RPC) mechanism or Remote Method Invocation (RMI) mechanism. Web Services and REST API are examples of object-oriented architecture. Invocations of methods are how objects communicate with one another. Typically, these are referred to as Remote Procedure Calls (RPC). REST API Calls, Web Services, and Java RMI are a few well-known examples. These characteristics apply to this.

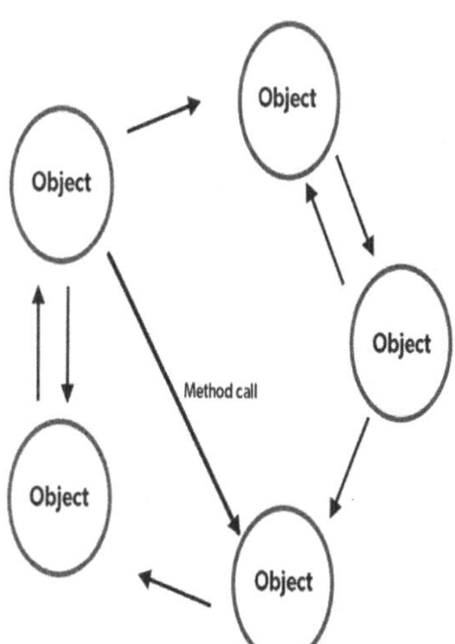

3. Data Centered Architecture:

Data Centered Architecture is a type of architecture in which a common data space is present at the centre. It contains all the required data in one place a shared data space. All the components are connected to this data space and they follow publish/subscribe type of communication. It has a central data repository at the centre. Required data is then delivered to the components. Distributed file systems, producer-consumer systems, and web-based data services are a few well-known examples.

For example Producer-Consumer system. The producer produces data in common data space and consumers request data.

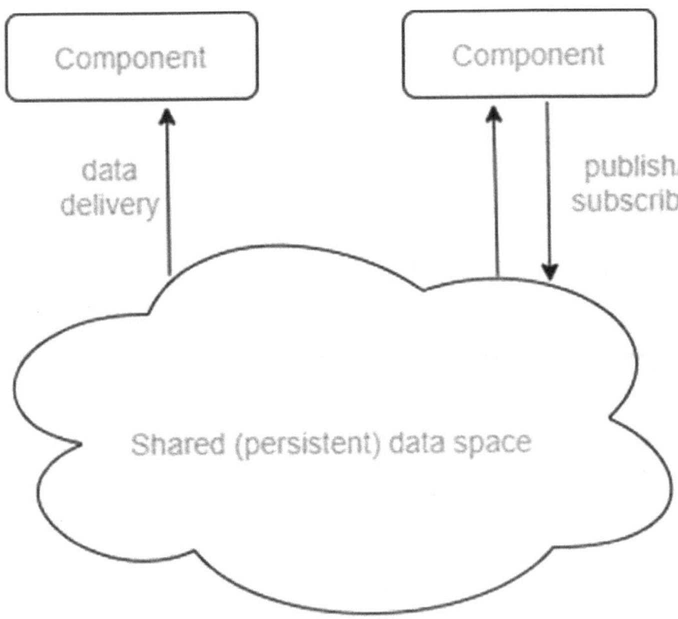

4. Event-Based Architecture:

Event-Based Architecture is almost similar to Data centered architecture just the difference is that in this architecture events are present instead of data. Events are present at the centre in the Event bus and delivered to the required component whenever needed. In this architecture, the entire communication is done through events. When an event occurs, the system, as well as the receiver, get notified. Data, URLs etc are transmitted through events. The components of this system are loosely coupled that's why it is easy to add, remove and modify them. Heterogeneous components can communicate through the bus. One significant benefit is that these heterogeneous components can communicate with the bus using any protocol.

However, a specific bus or an ESB has the ability to handle any kind of incoming request and respond appropriately.

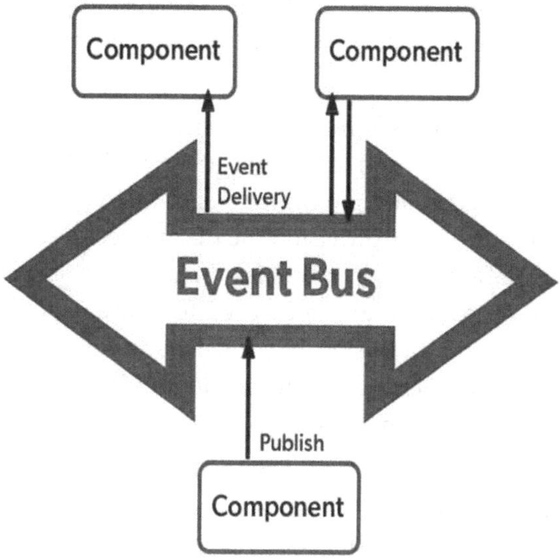

Types of Distributed Computing System Models

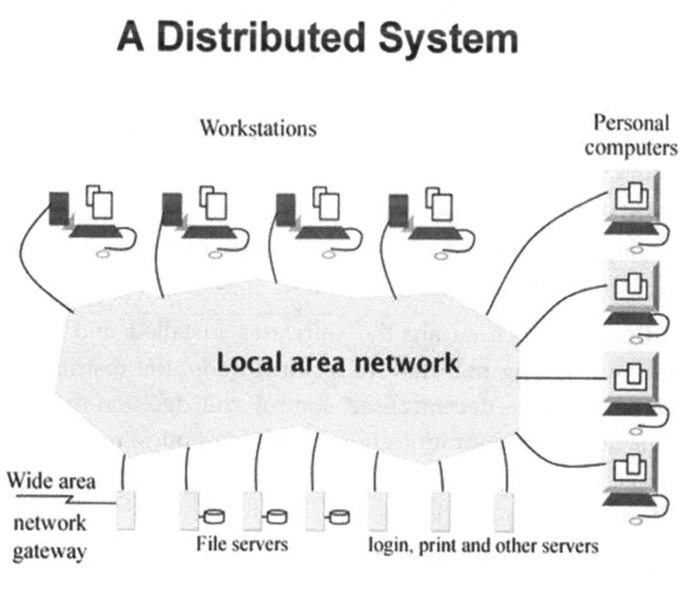

Physical Model

A physical model is basically a representation of the underlying hardware elements of a distributed system. It encompasses the hardware composition of a distributed system in terms of computers and other devices and their interconnections. It is primarily used to design, manage, implement and determine the performance of a distributed system. A physical model majorly consists of the following components:

- **Nodes** – Nodes are the end devices that have the ability of processing data, executing tasks and communicating with the other nodes. These end devices are generally the computers at the user end or can be servers, workstations etc. Nodes provision the distributed system with an interface in the presentation layer that enables the user to interact with other back-end devices, or nodes, that can be used for storage and database services, or processing, web browsing etc. Each node has an Operating System, execution environment and different middleware requirements that facilitate communication and other vital tasks.
- **Links** – Links are the communication channels between different nodes and intermediate devices. These may be wired or wireless. Wired links or physical media are implemented using copper wires, fibre optic cables etc. The choice of the medium depends on the environmental conditions and the requirements. Generally, physical links are required for high performance and real-time computing. Different connection types that can be implemented are as follows:
    - **Point-to-point links** – It establishes a connection and allows data transfer between only two nodes.
    - **Broadcast links** – It enables a single node to transmit data to multiple nodes simultaneously.
    - **Multi-Access links** – Multiple nodes share the same communication channel to transfer data. Requires protocols to avoid interference while transmission.
- **Middleware** – These are the softwares installed and executed on the nodes. By running middleware on each node, the distributed computing system achieves a decentralised control and decision-making. It handles various tasks like communication with other nodes, resource management, fault tolerance, synchronisation of different nodes and security to prevent malicious and unauthorised access.
- **Network Topology** – This defines the arrangement of nodes and links in the distributed computing system. The most common network topologies that are implemented are bus, star, mesh, ring or hybrid. Choice of topology is done by determining the exact use cases and the requirements.

- **Communication Protocols –** Communication protocols are the set rules and procedures for transmitting data from in the links. Examples of these protocols include TCP, UDP, HTTPS, MQTT etc. These allow the nodes to communicate and interpret the data.

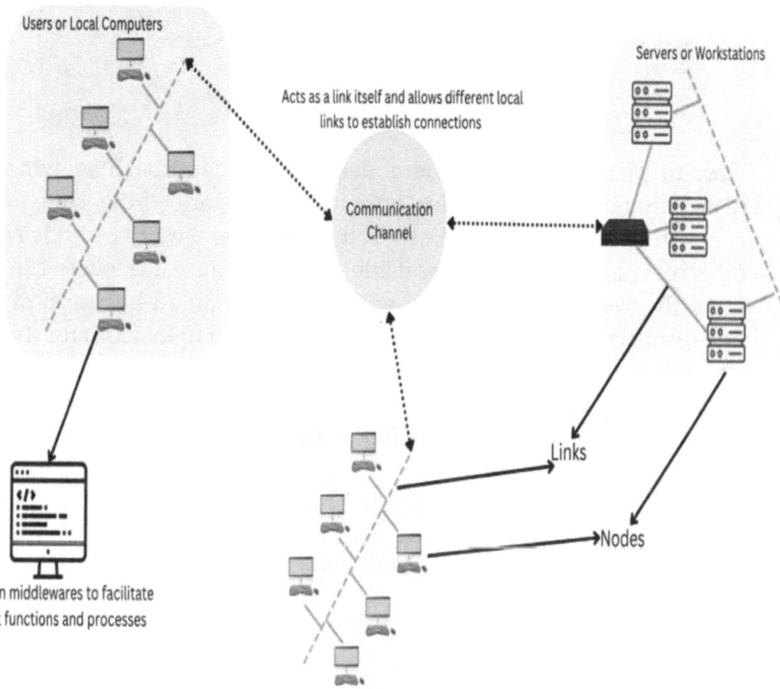

Architectural Model

Architectural model in distributed computing system is the overall design and structure of the system, and how its different components are organised to interact with each other and provide the desired functionalities. It is an overview of the system, on how will the development, deployment and operations take place. Construction of a good architectural model is required for efficient cost usage, and highly improved scalability of the applications. The key aspects of architectural model are –

- **Client-Server model –** It is a centralised approach in which the clients initiate requests for services and severs respond by providing those services. It mainly works on the request-response model where the client sends a request to the server and the server processes it, and responds to the client accordingly. It can be achieved by using TCP/IP, HTTP protocols on the transport layer. This is mainly used in web services, cloud computing, database management systems etc.

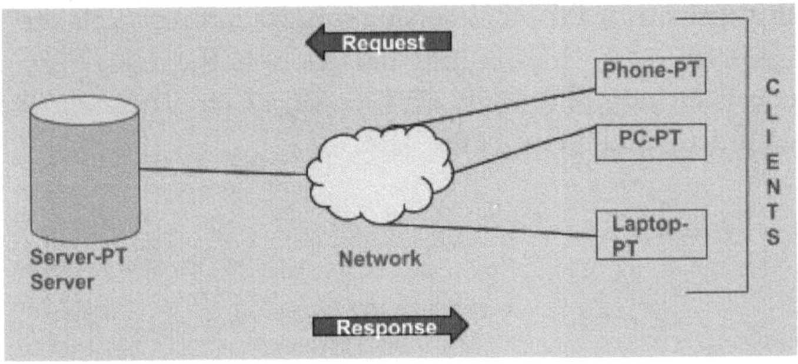

- **Peer-to-peer model** – It is a decentralised approach in which all the distributed computing nodes, known as peers, are all the same in terms of computing capabilities and can both request as well as provide services to other peers. It is a highly scalable model because the peers can join and leave the system dynamically, which makes it an ad-hoc form of network. The resources are distributed and the peers need to look out for the required resources as and when required. The communication is directly done amongst the peers without any intermediaries according to some set rules and procedures defined in the P2P networks. The best example of this type of computing is BitTorrent.

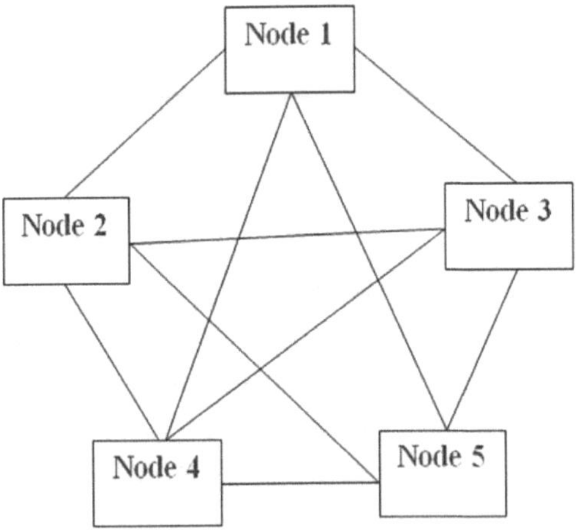

P2P Architecture

- **Layered model** – It involves organising the system into multiple layers, where each layer will provision a specific service. Each layer communicated with the adjacent layers using certain well-defined protocols without affecting the integrity of the system. A hierarchical structure is obtained where each layer abstracts the underlying complexity of lower layers.

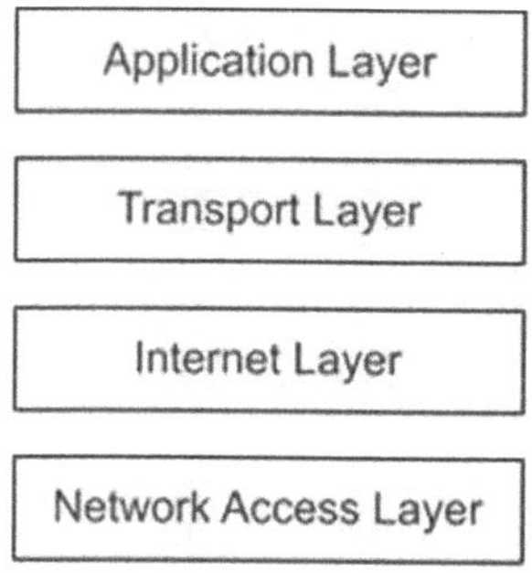

Various Layers of the TCP/ IP Model

- **Micro-services model –** In this system, a complex application or task, is decomposed into multiple independent tasks and these services running on different servers. Each service performs only a single function and is focussed on a specific business-capability. This makes the overall system more maintainable, scalable and easier to understand. Services can be independently developed, deployed and scaled without affecting the ongoing services.

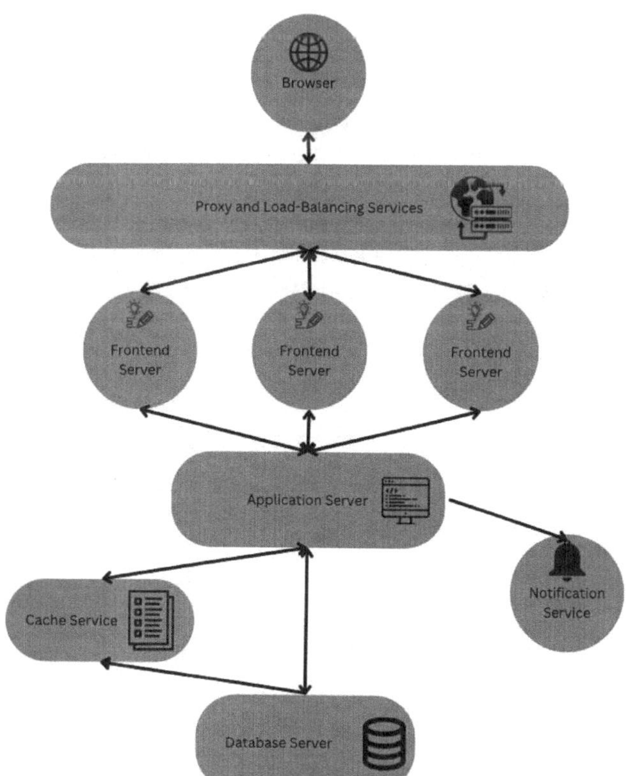

Fundamental Model

The fundamental model in a distributed computing system is a broad conceptual framework that helps in understanding the key aspects of the distributed systems. These are concerned with more formal description of properties that are generally common in all architectural models. It represents the essential components that are required to understand a distributed system's behaviour. Three fundamental models are as follows:

- **Interaction Model** – Distributed computing systems are full of many processes interacting with each other in highly complex ways. Interaction model provides a framework to understand the mechanisms and patterns that are used for communication and coordination among various processes. Different components that are important in this model are –
  - **Message Passing –** It deals with passing messages that may contain, data, instructions, a service request, or process synchronisation between different computing nodes. It may be synchronous or asynchronous depending on the types of tasks and processes.
  - **Publish/Subscribe Systems –** Also known as pub/sub system. In this the publishing process can publish a message over a topic and the processes that are subscribed to that topic can take it up and execute the process for themselves. It is more important in an event-driven architecture.
- **Remote Procedure Call (RPC) –** It is a communication paradigm that has an ability to invoke a new process or a method on a remote process as if it were a local procedure call. The client process makes a procedure call using RPC and then the message is passed to the required server process using communication protocols. These message passing protocols are abstracted and the result once obtained from the server process, is sent back to the client process to continue execution.

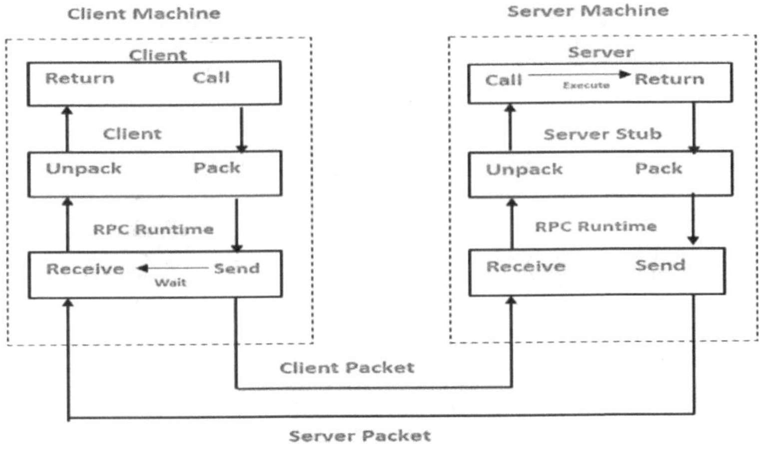

- **Failure Model** – This model addresses the faults and failures that occur in the distributed computing system. It provides a framework to identify and rectify the faults that occur or may occur in the system. Fault tolerance mechanisms are implemented so as to handle failures by replication and error detection and recovery methods. Different failures that may occur are:
    - **Crash failures** – A process or node unexpectedly stops functioning.
    - **Omission failures** – It involves a loss of message, resulting in absence of required communication.
    - **Timing failures** – The process deviates from its expected time quantum and may lead to delays or unsynchronised response times.
    - **Byzantine failures** – The process may send malicious or unexpected messages that conflict with the set protocols.
- **Security Model** – Distributed computing systems may suffer malicious attacks, unauthorised access and data breaches. Security model provides a framework for understanding the security requirements, threats, vulnerabilities, and mechanisms to safeguard the system and its resources. Various aspects that are vital in the security model are –
    - **Authentication** – It verifies the identity of the users accessing the system. It ensures that only the authorised and trusted entities get access. It involves –
        - **Password-based authentication** – Users provide a unique password to prove their identity.
        - **Public-key cryptography** – Entities possess a private key and a corresponding public key, allowing verification of their authenticity.
        - **Multi-factor authentication** – Multiple factors, such as passwords, biometrics, or security tokens, are used to validate identity.
    - **Encryption** – It is the process of transforming data into a format that is unreadable without a decryption key. It protects sensitive information from unauthorized access or disclosure.

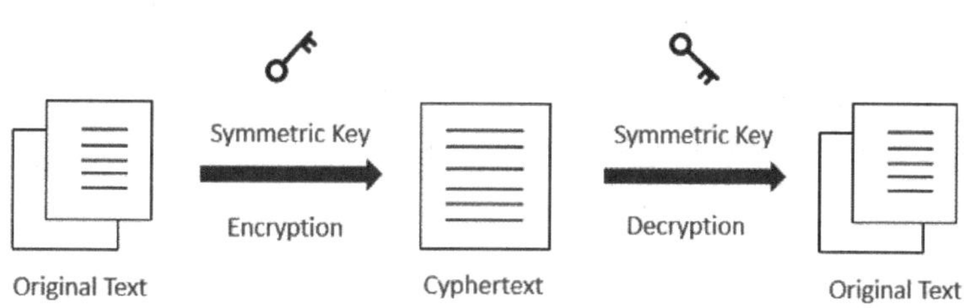

- **Data Integrity –** Data integrity mechanisms protect against unauthorised modifications or tampering of data. They ensure that data remains unchanged during storage, transmission, or processing. Data integrity mechanisms include:
    - **Hash functions** – Generating a hash value or checksum from data to verify its integrity.
    - **Digital signatures** – Using cryptographic techniques to sign data and verify its authenticity and integrity.

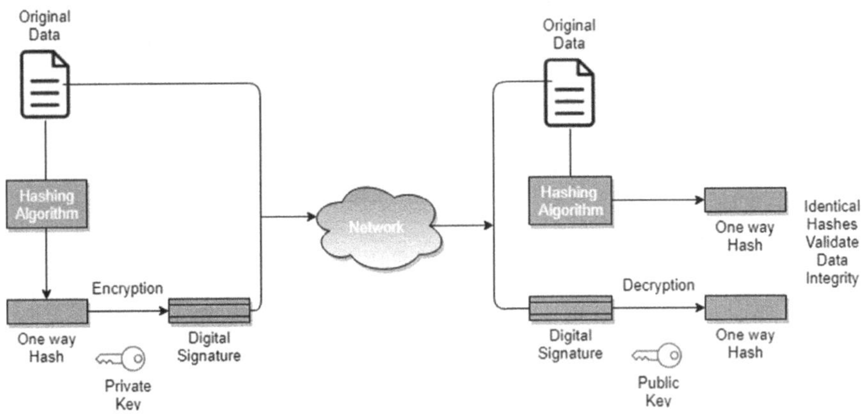

Cloud Computing

Simply put, cloud computing is the delivery of computing services—including servers, storage, databases, networking, software, analytics, and intelligence—over the internet ("the cloud") to offer faster innovation, flexible resources, and economies of scale. You typically pay only for cloud services you use, helping you lower your operating costs, run your infrastructure more efficiently, and scale as your business needs change.

Difference between Cloud Computing and Distributed Computing :

S.No.	CLOUD COMPUTING	DISTRIBUTED COMPUTING
01.	Cloud computing refers to providing on demand IT resources/services like server, storage, database, networking, analytics, software etc. over internet.	Distributed computing refers to solve a problem over distributed autonomous computers and they communicate between them over a network.

02.	In simple cloud computing can be said as a computing technique that delivers hosted services over the internet to its users/customers.	In simple distributed computing can be said as a computing technique which allows to multiple computers to communicate and work to solve a single problem.
03.	It is classified into 4 different types such as Public Cloud, Private Cloud, Community Cloud and Hybrid Cloud.	It is classified into 3 different types such as Distributed Computing Systems, Distributed Information Systems and Distributed Pervasive Systems.
04.	There are many benefits of cloud computing like cost effective, elasticity and reliable, economies of Scale, access to the global market etc.	There are many benefits of distributed computing like flexibility, reliability, improved performance etc.
05.	Cloud computing provides services such as hardware, software, networking resources through internet.	Distributed computing helps to achieve computational tasks more faster than using a single computer as it takes a lot of time.
06.	**The goal of cloud computing is to provide on demand computing services over internet on pay per use model.**	**The goal of distributed computing is to distribute a single task among multiple computers and to solve it quickly by maintaining coordination between them.**
07.	**Some characteristics of cloud computing are providing shared pool of configurable computing resources, on-demand service, pay per use, provisioned by the Service Providers etc.**	**Some characteristics of distributed computing are distributing a single task among computers to progress the work at same time, Remote Procedure calls and Remote Method Invocation for distributed computations.**
08.	**Some disadvantage of cloud computing includes less control especially in the case of public clouds, restrictions on available services may be faced and cloud security.**	**Some disadvantage of distributed computing includes chances of failure of nodes, slow network may create problem in communication.**

Principles to Scale Up Cloud Computing

This section will discuss the principles that leverage the Internet to scale up cloud computing services.

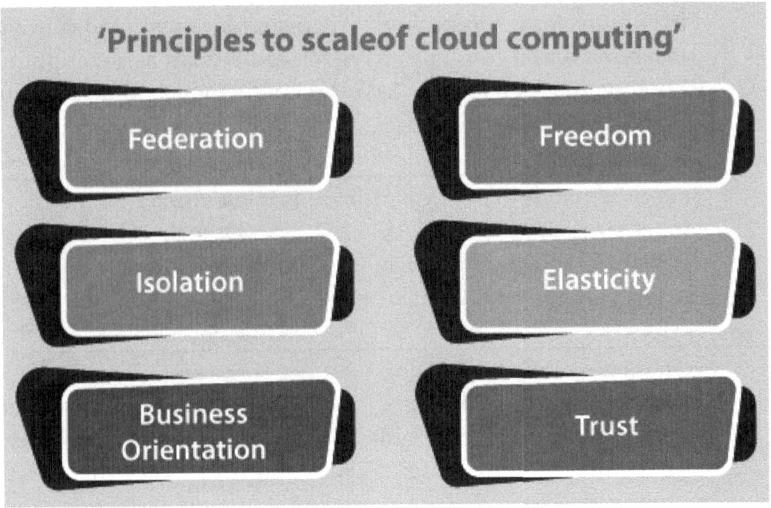

1. Federation

Cloud resources are always unlimited for customers, but each cloud has a limited capacity. If customer demand continues to grow, the cloud will have to exceed its potential, for which the form federation of service providers enables collaboration and resource sharing.

A federated cloud must allow virtual applications to be deployed on federated sites. Virtual applications should not be location-dependent and should be able to migrate easily between sites.

Union members should be independent, making it easier for competing service providers to form unions.

2. Freedom

Cloud computing services should provide end-users complete freedom that allows the user to use cloud services without depending on a specific cloud provider.

Even the cloud provider should be able to manage and control the computing service without sharing internal details with customers or partners.

3. Isolation

We are all aware that a cloud service provider provides its computing resources to multiple end-users. The end-user must be assured before moving his computing cloud that his data or information will be isolated in the cloud and cannot be accessed by other members sharing the cloud.

4. Elasticity

Cloud computing resources should be elastic, which means that the user should be free to attach and release computing resources on their demand.

5. Business Orientation

Companies must ensure the quality of service providers offer before moving mission-critical applications to the cloud. The cloud service provider should develop a mechanism to understand the exact business requirement of the customer and customize the service parameters as per the customer's requirement.

6. Trust

Trust is the most important factor that drives any customer to move their computing to the cloud. For the cloud to be successful, trust must be maintained to create a federation between the cloud customer, the cloud vendor, and the various cloud providers.

So, these are the principles of cloud computing that take advantage of the Internet to enhance cloud computing. A cloud provider considers these principles before deploying cloud services to end-users.

What is cloud architecture?

Cloud architecture is a key element of building in the cloud. It refers to the layout and connects all the necessary components and technologies required for cloud computing.

Migrating to the cloud can offer many business benefits compared to on-premises environments, from improved agility and scalability to cost efficiency. While many organizations may start with a "lift-and-shift" approach, where on-premises applications are moved over with minimal modifications, ultimately it will be necessary to construct and deploy applications according to the needs and requirements of cloud environments.

Cloud architecture dictates how components are integrated so that you can pool, share, and scale resources over a network. Think of it as a building blueprint for running and deploying applications in cloud environments.

Explore how Google Cloud helps you design cloud architecture to match your business needs. Use our Architecture Framework for guidance, recommendations, and best practices to build and migrate your workloads to the cloud. Use our Architecture Diagramming Tool for pre-built reference architectures and customizing them to your use cases.

Cloud architecture defined

Cloud architecture refers to how various cloud technology components, such as hardware, virtual resources, software capabilities, and virtual network systems interact and connect to create cloud computing environments. It acts as a blueprint that defines the best way to strategically combine resources to build a cloud environment for a specific business need.

Cloud architecture components

Cloud architecture components include:

- A frontend platform
- A backend platform
- A cloud-based delivery model
- A network (internet, intranet, or intercloud)

In cloud computing, frontend platforms contain the client infrastructure—user interfaces, client-side applications, and the client device or network that enables users to interact with and access cloud computing services. For example, you can open the web browser on your mobile phone and edit a Google Doc. All three of these things describe frontend cloud architecture components.

On the other hand, the back end refers to the cloud architecture components that make up the cloud itself, including computing resources, storage, security mechanisms, management, and more.

Below is a list of the main backend components:

Application: The backend software or application the client is accessing from the front end to coordinate or fulfill client requests and requirements.

Service: The service is the heart of cloud architecture, taking care of all the tasks being run on a cloud computing system. It manages which resources you can access, including storage, application development environments, and web applications.

Runtime cloud: Runtime cloud provides the environment where services are run, acting as an operating system that handles the execution of service tasks and management. Runtimes use virtualization technology to create hypervisors that represent all your services, including apps, servers, storage, and networking.

Storage: The storage component in the back end is where data to operate applications is stored. While cloud storage options vary by provider, most cloud service providers offer flexible scalable storage services that are designed to store and manage vast amounts of data in the cloud. Storage may include hard drives, solid-state drives, or persistent disks in server bays.

Infrastructure: Infrastructure is probably the most commonly known component of cloud architecture. In fact, you might have thought that cloud infrastructure *is* cloud architecture. However, cloud infrastructure comprises all the major hardware components that power cloud services, including the CPU, graphics processing unit (GPU), network devices, and other hardware components needed for systems to run smoothly. Infrastructure also refers to all the software needed to run and manage everything.

Cloud architecture, on the other hand, is the plan that dictates how cloud resources and infrastructure are organized.

Management: Cloud service models require that resources be managed in real time according to user requirements. It is essential to use management software, also known as middleware, to coordinate communication between the backend and frontend cloud architecture components and allocate resources for specific tasks. Beyond middleware, management software will also include capabilities for usage monitoring, data integration, application deployment, and disaster recovery.

Security: As more organizations continue to adopt cloud computing, implementing cloud security features and tools is critical to securing data, applications, and platforms. It's essential to plan and design data security and network security to provide visibility, prevent data loss and downtime, and ensure redundancy. This may include regular backups, debugging, and virtual firewalls.

How does cloud architecture work?

In cloud architecture, each of the components works together to create a cloud computing platform that provides users with on-demand access to resources and services.

The back end contains all the cloud computing resources, services, data storage, and applications offered by a cloud service provider. A network is used to connect the frontend and backend cloud architecture components, enabling data to be sent back and forth between them. When users interact with the front end (or client-side interface), it sends queries to the back end using middleware where the service model carries out the specific task or request.

The types of services available to use vary depending on the cloud-based delivery model or service model you have chosen. There are three main cloud computing service models:

- **Infrastructure as a service (IaaS):** This model provides on-demand access to cloud infrastructure, such as servers, storage, and networking. This eliminates the need to procure, manage, and maintain on-premises infrastructure.
- **Platform as a service (PaaS):** This model offers a computing platform with all the underlying infrastructure and software tools needed to develop, run, and manage applications.
- **Software as a service (SaaS):** This model offers cloud-based applications that are delivered and maintained by the service provider, eliminating the need for end users to deploy software locally.

Cloud architecture layers

A simpler way of understanding how cloud architecture works is to think of all these components as various layers placed on top of each other to create a cloud platform.

Here are the basic cloud architecture layers:

1. **Hardware:** The servers, storage, network devices, and other hardware that power the cloud.
2. **Virtualization:** An abstraction layer that creates a virtual representation of physical computing and storage resources. This allows multiple applications to use the same resources.
3. **Application and service:** This layer coordinates and supports requests from the frontend user interface, offering different services based on the cloud service model, from resource allocation to application development tools to web-based applications.

Types of cloud architecture

Cloud adoption is not one-size-fits-all. You'll need to consider what type of cloud you want to build based on your existing technology investments, your specific business requirements, and the overall goals you hope to achieve.

There are three main types of cloud architecture you can choose from: **public, private, and hybrid**.

Public cloud architecture uses cloud computing resources and physical infrastructure that is owned and operated by a third-party cloud service provider. Public clouds enable you to scale resources easily without having to invest in your own hardware or software, but use multi-tenant architectures that serve other customers at the same time.

Private cloud architecture refers to a dedicated cloud that is owned and managed by your organization. It is privately hosted on-premises in your own data center, providing more control over resources and more security over data and infrastructure. However, this architecture is considerably more expensive and requires more IT expertise to maintain.

Hybrid cloud architecture uses both public and private cloud architecture to deliver a flexible mix of cloud services. A hybrid cloud allows you to migrate workloads between environments, allowing you to use the services that best suit your business demands and the workload. Hybrid cloud architectures are often the solution of choice for businesses that need control over their data but also want to take advantage of public cloud offerings.

In recent years, **multicloud architecture** is also emerging as more organizations look to use cloud services from multiple cloud providers. Multicloud environments are gaining popularity for their flexibility and ability to better match use cases to specific offerings, regardless of vendor.

What does a cloud architect do?

A cloud architect is an IT expert responsible for developing, implementing, and managing an organization's cloud architecture. As cloud strategies continue to become more complex, the skills and expertise of cloud architects are becoming more vital for helping companies navigate the complexities of cloud environments, implement successful strategies, and keep cloud systems running smoothly.

Benefits of cloud architecture

There are many benefits of cloud architecture for organizations, including:

Cost-effective

Instead of investing upfront costs for servers, you can opt to use the infrastructure of a cloud service provider. Dynamic provisioning allows you to further optimize spending by paying only for the computing resources you use.

Faster time to market

You no longer need to wait to procure, set up, and configure computing infrastructure. Cloud architectures enable you to get up and running fast, so you spend more time focusing on developing and delivering new products.

Scalability

Cloud architectures give you more flexibility to scale computing resources up (or down) based on your infrastructure requirements. You can easily scale to meet higher demand, whether from growth or seasonal spikes in traffic.

Accelerated transformation

Cloud-native architectures like Kubernetes let you make the most of cloud services and automated environments to speed up modernization and drive digital transformation.

More innovation

Cloud architectures allow you to leverage the latest technologies for storage, security, analytics, and AI like machine learning.

High availability

Applications run and managed on cloud architectures benefit from high-performance computing resources that ensure continuous availability, regardless of fluctuating loads.

Strong security

Cloud service providers consistently upgrade and improve their security mechanisms with expert professionals and the latest technologies to help secure your data, systems, and workloads.

Cloud Computing , which is one of the demanding technology of the current time and which is giving a new shape to every organization by providing on demand virtualized services/resources. Starting from small to medium and medium to large, every organization use cloud computing services for storing information and accessing it from anywhere and any time only with the help of internet. In this article, we will know more about the internal architecture of cloud computing.

Transparency, scalability, security and intelligent monitoring are some of the most important constraints which every cloud infrastructure should experience. Current research on other important constraints is helping cloud computing system to come up with new features and strategies with a great capability of providing more advanced cloud solutions.

Cloud Computing Architecture :

The cloud architecture is divided into 2 parts i.e.

1. Frontend
2. Backend

The below figure represents an internal architectural view of cloud computing.

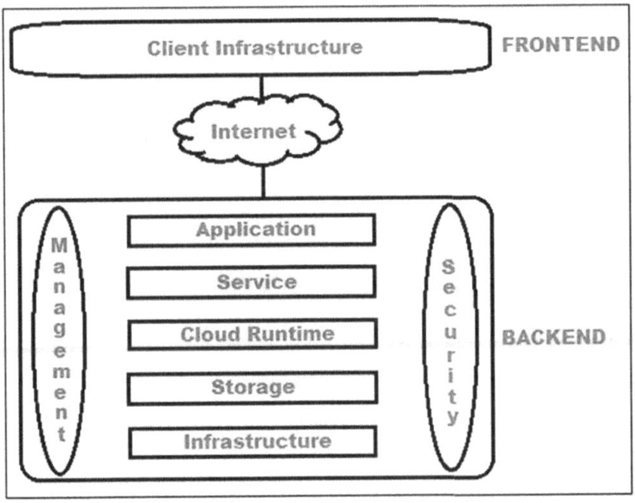

Architecture of cloud computing is the combination of both SOA (Service Oriented Architecture) and EDA (Event Driven Architecture). Client infrastructure, application, service, runtime cloud, storage, infrastructure, management and security all these are the components of cloud computing architecture.

1. Frontend :

Frontend of the cloud architecture refers to the client side of cloud computing system. Means it contains all the user interfaces and applications which are used by the client to access the cloud computing services/resources. For example, use of a web browser to access the cloud platform.

- **Client Infrastructure** – Client Infrastructure is a part of the frontend component. It contains the applications and user interfaces which are required to access the cloud platform.
- In other words, it provides a GUI(Graphical User Interface) to interact with the cloud.

2. Backend :

Backend refers to the cloud itself which is used by the service provider. It contains the resources as well as manages the resources and provides security mechanisms. Along with this, it includes huge storage, virtual applications, virtual machines, traffic control mechanisms, deployment models, etc.

1. Application –

Application in backend refers to a software or platform to which client accesses. Means it provides the service in backend as per the client requirement.

2. Service –

Service in backend refers to the major three types of cloud based services like SaaS, PaaS and IaaS. Also manages which type of service the user accesses.

3. Runtime Cloud-

Runtime cloud in backend provides the execution and Runtime platform/ environment to the Virtual machine.

4. Storage –

Storage in backend provides flexible and scalable storage service and management of stored data.

5. Infrastructure –

Cloud Infrastructure in backend refers to the hardware and software components of cloud like it includes servers, storage, network devices, virtualization software etc.

6. Management –

Management in backend refers to management of backend components like application, service, runtime cloud, storage, infrastructure, and other security mechanisms etc.

7. Security –

Security in backend refers to implementation of different security mechanisms in the backend for secure cloud resources, systems, files, and infrastructure to end-users.

8. Internet –

Internet connection acts as the medium or a bridge between frontend and backend and establishes the interaction and communication between frontend and backend.

9. Database–

Database in backend refers to provide database for storing structured data, such as SQL and NOSQL databases. Example of Databases services include Amazon RDS, Microsoft Azure SQL database and Google CLoud SQL.

10. Networking–

Networking in backend services that provide networking infrastructure for application in the cloud, such as load balancing, DNS and virtual private networks.

11. Analytics–

Analytics in backend service that provides analytics capabillities for data in the cloud, such as warehousing, bussness intellegence and machine learning.

Benefits of Cloud Computing Architecture :

- Makes overall cloud computing system simpler.
- Improves data processing requirements.
- Helps in providing high security.
- Makes it more modularized.
- Results in better disaster recovery.
- Gives good user accessibility.
- Reduces IT operating costs.
- Provides high level reliability.
- Scalability.

There are four fundamentals you should know when designing and building cloud environments:

- Cloud infrastructure components
- Cloud delivery model
- Cloud service model (Cloud infrastructure as a service)
- Cloud deployment model (types of cloud architectures)

Let's cover each briefly:

1. Cloud infrastructure components

The following are the building materials or the fundamentals of cloud architecture:

- Cloud infrastructure – Similar to traditional computing, this refers to computing power (servers), storage (hard drives and flash drives in a data center), and networking (routers and switches).
- Virtualization – Virtual representations of physical servers, networks, and storage in cloud environments create an abstraction layer from which multiple applications can run. A cloud environment can thus offer near-

limitless resources (scalability). You can select the ideal resources for your business (agility), such as CPUs, RAMs, and storage types. You can also adjust these resources as your needs change (flexibility).

- Middleware – This comprises all components that facilitate communication between networked infrastructures, applications, and services.
- Management – This refers to governing your cloud environment through a central dashboard to maximize performance while limiting downtime.
- Automation software – Software assists in cloud management. It allows you to allocate the right amount of resources despite fluctuating demand to ensure that your cloud performs optimally at all times. As with virtualization, this is yet another key difference between cloud and on-premise architecture.

Cloud deployment models allow clients to access the cloud's resources.

2. Cloud delivery models

Here, cloud architecture is two-fold; frontend and backend. The front end is client-facing and comprises the user interfaces and applications that a client uses to access cloud computing resources.

Computing resources are in the backend and include applications, services, cloud runtime, and storage. Some clouds allow clients to connect directly with "bare metal" servers.

Now, we did mention that architecting a cloud computing environment is simpler than building a house from scratch. Here is why.

3. Cloud service model

Building a home requires a lot of site inspections before you can lay the building's foundation. It starts from scratch, from leveling the ground and pouring the foundation to completing rough framing and installing plumbing.

Building on cloud architecture does not have to be a start-from-scratch project.

You can begin quickly and affordably using a cloud service provider's infrastructure. Cloud service providers offer this subscription service in three principal models:

- An infrastructure-as-a-service (IaaS) model allows your organization to rent scalable server space from a cloud provider monthly. While the vendor provides the hardware, you handle your environment's applications, middleware, and configuration.
- A platform-as-a-service (PaaS) is when the vendor assists in more than just virtualizing infrastructure but also offers cloud infrastructure management and a computing platform for testing applications.
- With Software-as-a-Service (SaaS), your organization can use pre-built applications and software via remote servers, either immediately through web interfaces or after customizing it via APIs.

4. Types of cloud architectures (Cloud deployment models)

Here are the dominant four you should know:

- Public cloud architecture involves a single cloud provider servicing the hardware needs of several organizations. As such, public clouds have multi-tenant architectures compared to single-tenant architectures to serve many customers concurrently and effectively.
- With private cloud architecture, your organization develops an ideal cloud service model internally. You dedicate it to the needs of just one tenant, so it is private.
- A hybrid cloud architecture refers to an organization adopting aspects of both public and private cloud architectures to optimize costs and performance. The architecture uses VPN links to switch workloads between public and private clouds.
- A multi-cloud architecture lets an organization take advantage of two or more cloud providers' services to meet regulatory requirements, maximize vendor lock-in, and use best-in-class solutions.
- Cloud-native architecture is an application development approach built on the principles of distributed workloads, microservices, containers, and serverless functions. It emphasizes building from scratch or refactoring legacy applications to take full advantage of cloud infrastructure.

NIST Cloud Computing Reference Architecture Cloud models:

NIST Cloud Computing reference architecture defines five major performers:

- Cloud Provider
- Cloud Carrier
- Cloud Broker
- Cloud Auditor
- Cloud Consumer

Each performer is an object (a person or an organization) that contributes to a transaction or method and/or performs tasks in Cloud computing. There are five major actors defined in the NIST cloud computing reference architecture, which are described below:

1. Cloud Service Providers:

A group or object that delivers cloud services to cloud consumers or end-users. It offers various components of cloud computing. Cloud computing consumers purchase a growing variety of cloud services from cloud service providers. There are various categories of cloud-based services mentioned below:

- **IaaS Providers:** In this model, the cloud service providers offer infrastructure components that would exist in an on-premises data center. These components consist of servers, networking, and storage as well as the virtualization layer.

- **SaaS Providers:** In Software as a Service (SaaS), vendors provide a wide sequence of business technologies, such as Human resources management (HRM) software, customer relationship management (CRM) software, all of which the SaaS vendor hosts and provides services through the internet.
- **PaaS Providers:** In Platform as a Service (PaaS), vendors offer cloud infrastructure and services that can access to perform many functions. In PaaS, services and products are mostly utilized in software development. PaaS providers offer more services than IaaS providers. PaaS providers provide operating system and middleware along with application stack, to the underlying infrastructure.

2. Cloud Carrier: The mediator who provides offers connectivity and transport of cloud services within cloud service providers and cloud consumers. It allows access to the services of the cloud through Internet networks, telecommunication, and other access devices. Network and telecom carriers or a transport agent can provide distribution. A consistent level of services is provided when cloud providers set up Service Level Agreements (SLA) with a cloud carrier. In general, Carrier may be required to offer dedicated and encrypted connections.

3. Cloud Broker: An organization or a unit that manages the performance, use, and delivery of cloud services by enhancing specific capability and offers value-added services to cloud consumers. It combines and integrates various services into one or more new services. They provide service arbitrage which allows flexibility and opportunistic choices. There are major three services offered by a cloud broker:

- Service Intermediation.
- Service Aggregation.
- Service Arbitrage.

4. Cloud Auditor: An entity that can conduct independent assessment of cloud services, security, performance, and information system operations of the cloud implementations. The services that are provided by Cloud Service Providers (CSP) can be evaluated by service auditors in terms of privacy impact, security control, and performance, etc. Cloud Auditor can make an assessment of the security controls in the information system to determine the extent to which the controls are implemented correctly, operating as planned and constructing the desired outcome with respect to meeting the security necessities for the system. There are three major roles of Cloud Auditor which are mentioned below:

- Security Audit.
- Privacy Impact Audit.
- Performance Audit.

5. Cloud Consumer: A cloud consumer is the end-user who browses or utilizes the services provided by Cloud Service Providers (CSP), sets up service contracts with the cloud provider. The cloud consumer pays per use of the service provisioned. Measured services utilized by the consumer. In this, a set of organizations having mutual regulatory constraints performs a security and risk assessment for each use case of Cloud migrations and deployments. Cloud consumers use Service-Level Agreement (SLAs) to specify the technical performance requirements to be fulfilled by a cloud provider. SLAs can cover terms concerning the quality of service, security, and remedies for performance failures. A cloud provider may also list in the SLAs a set of limitations or boundaries, and obligations that cloud consumers must accept. In a mature market environment, a cloud consumer can freely pick a cloud provider with better pricing and more favourable terms. Typically, a cloud provider's public pricing policy and SLAs are non-negotiable, although a cloud consumer who assumes to have substantial usage might be able to negotiate for better contracts.

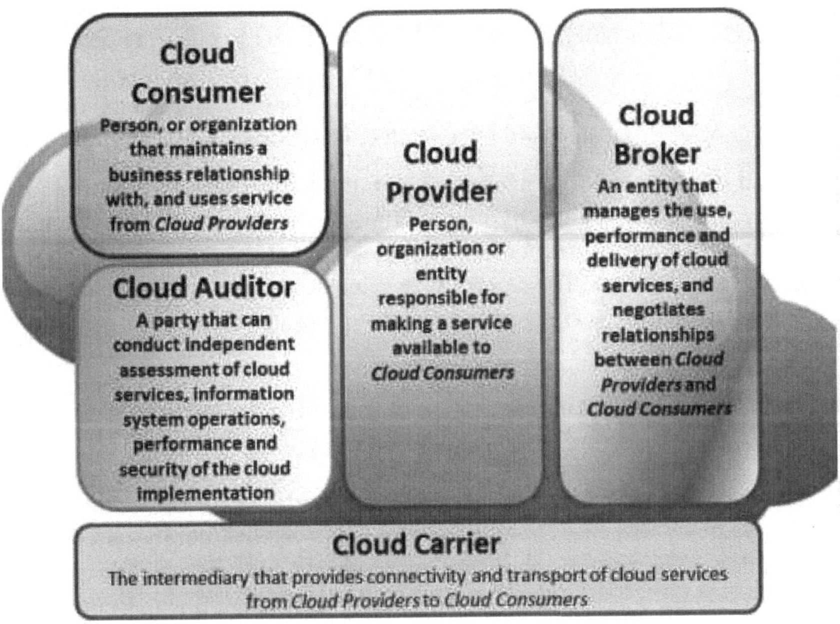

Fig. Cloud Actors

Cloud Service Models

There are the following three types of cloud service models -

1. Infrastructure as a Service (IaaS)
2. Platform as a Service (PaaS)
3. Software as a Service (SaaS)

Infrastructure as a Service (IaaS)

IaaS is also known as **Hardware as a Service (HaaS)**. It is a computing infrastructure managed over the internet. The main advantage of using IaaS is that it helps users to avoid the cost and complexity of purchasing and managing the physical servers.

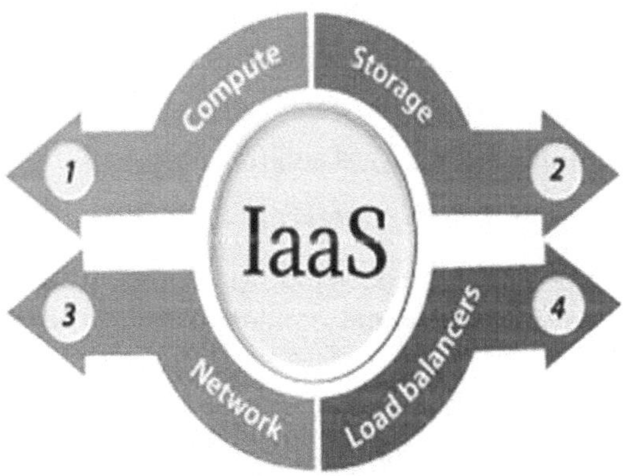

Characteristics of IaaS

There are the following characteristics of IaaS -

- Resources are available as a service
- Services are highly scalable
- Dynamic and flexible
- GUI and API-based access
- Automated administrative tasks

Example: DigitalOcean, Linode, Amazon Web Services (AWS), Microsoft Azure, Google Compute Engine (GCE), Rackspace, and Cisco Metacloud.

Platform as a Service (PaaS)

PaaS cloud computing platform is created for the programmer to develop, test, run, and manage the applications.

Characteristics of PaaS

There are the following characteristics of PaaS -

- Accessible to various users via the same development application.
- Integrates with web services and databases.
- Builds on virtualization technology, so resources can easily be scaled up or down as per the organization's need.
- Support multiple languages and frameworks.
- Provides an ability to "**Auto-scale**".

Example: AWS Elastic Beanstalk, Windows Azure, Heroku, Force.com, Google App Engine, Apache Stratos, Magento Commerce Cloud, and OpenShift.

Software as a Service (SaaS)

SaaS is also known as "**on-demand software**". It is a software in which the applications are hosted by a cloud service provider. Users can access these applications with the help of internet connection and web browser.

Characteristics of SaaS

There are the following characteristics of SaaS -

- Managed from a central location
- Hosted on a remote server
- Accessible over the internet
- Users are not responsible for hardware and software updates. Updates are applied automatically.
- The services are purchased on the pay-as-per-use basis

Example: BigCommerce, Google Apps, Salesforce, Dropbox, ZenDesk, Cisco WebEx, ZenDesk, Slack, and GoToMeeting.

Difference between IaaS, PaaS, and SaaS

The below table shows the difference between IaaS, PaaS, and SaaS -

IaaS	Paas	SaaS
It provides a virtual data center to store information and create platforms for app development, testing, and deployment.	It provides virtual platforms and tools to create, test, and deploy apps.	It provides web software and apps to complete business tasks.
It provides access to resources such as virtual machines, virtual storage, etc.	It provides runtime environments and deployment tools for applications.	It provides software as a service to the end-users.
It is used by network architects.	It is used by developers.	It is used by end users.
IaaS provides only Infrastructure.	PaaS provides Infrastructure+Platform.	SaaS provides Infrastructure Platform+Software.

1. Public Cloud:

Computing in which service provider makes all resources public over the internet. It is connected to the public Internet. Service provider serves resources such as virtual machines, applications, storage, etc to the general public over the internet. It may be free of cost or with minimal pay-per-usage. It is available for public display, Google uses the cloud to run some of its applications like google docs, google drive orYouTube, etc.

It is the most common way of implementing cloud computing. The external cloud service provider owns, operates, and delivers it over the publicnetwork.

It is best for the companies which need an infrastructure to accommodate a large number of customers and work on projects which have diverse organizations i.e. research institutions and NGOs etc.

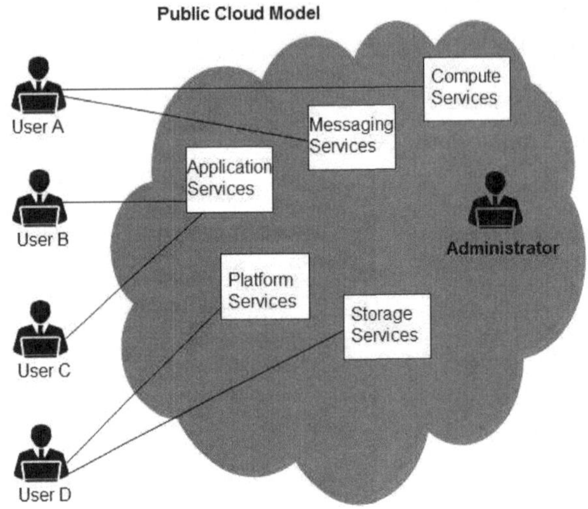

Public Cloud Advantages:

- **Scalability:** Public cloud services are designed to be scalable, which means you can easily add or remove resources as needed.
- **Cost-Effective:** Public cloud providers typically offer a pay-as-you-go model, which allows you to only pay for the resources you use, making it cost-effective.
- **Accessibility:** Public cloud services are accessible from anywhere with an internet connection, which makes it easy to access data and applications from anywhere in the world.
- **Reliability:** Public cloud providers usually have multiple data centers across different locations, which ensures high availability and reliability.

Public Cloud Disadvantages:

- **Security:** Since public cloud services are available to anyone with an internet connection, there is a risk of unauthorized access or data breaches.
- **Control:** Public cloud providers have control over the infrastructure and services, which means you have limited control over how your data is stored and managed.
- **Compliance:** Compliance requirements can be a challenge for public cloud users since they have to comply with the regulations and standards that apply to the cloud provider.

2. Private Cloud:

Computing in which service provider does not makes all resources public over the internet. It only supports connectivity over the private network. It has only authentic users and single-occupant architecture. Google back-end data of the applications like Google Drive, Google docs, YouTube, etc are not available to the public, these types of data and applications run on a private cloud.

The infrastructure and services are maintained and deployed over a private network; hardware and software are dedicated only to a private company i.e. members of the special entity.

It is best for the companies which need an infrastructure that has high performance, high security, and privacy due to its best adaptability and flexibility.

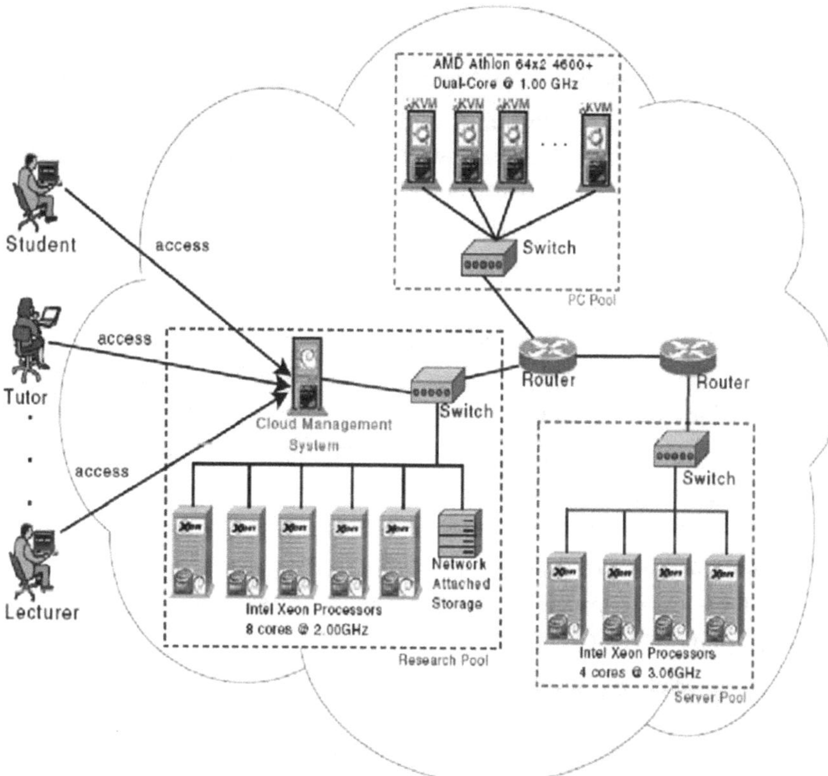

Private Cloud Advantages:

- **Control:** Private cloud users have full control over the infrastructure and services, which means they can customize the environment to their specific needs.
- **Security:** Private cloud services are typically more secure since they are only accessible to authorized users.
- **Compliance:** Private cloud users have more control over compliance requirements since they can customize the environment to meet specific regulatory requirements.
- **Customization:** Private cloud users can customize the environment to their specific needs, which means they can optimize performance and efficiency.

Private Cloud Disadvantages:

- **Cost:** Private cloud services can be more expensive than public cloud services since they require a dedicated infrastructure.
- **Maintenance:** Private cloud users are responsible for maintaining and updating the infrastructure, which can be time-consuming and costly.
- **Scalability:** Private cloud services can be less scalable than public cloud services, which means they may not be able to handle sudden spikes in demand.

Similarities between Public Cloud and Private Cloud, which include:

- **Virtualization:** Both Public Cloud and Private Cloud use virtualization technologies to create virtual machines, which enables them to maximize the use of physical hardware and resources.
- **Automation:** Both Public Cloud and Private Cloud use automation tools to manage and provision resources, which helps to streamline operations and improve efficiency.
- **High Availability:** Both Public Cloud and Private Cloud are designed to provide high availability and reliability, with multiple data centers, redundant systems, and failover mechanisms to minimize downtime.
- **Scalability:** Both Public Cloud and Private Cloud are designed to be scalable, which allows users to easily add or remove resources as needed to meet changing demands.
- **Resource Pooling:** Both Public Cloud and Private Cloud utilize resource pooling, where multiple users or applications share resources such as storage, networking, and computing power.
- **Service-oriented Architecture:** Both Public Cloud and Private Cloud use a service-oriented architecture (SOA), where resources are provided as services that can be accessed and used by multiple applications or users.

Difference Between Public Cloud and Private Cloud

Cloud computing is a way of providing IT infrastructure to customers, it is not just a set of products to be implemented. For any service to be a cloud service, the following five criteria need to be fulfilled as follows:

1. **On-demand self-service:** Decision of starting and stopping service depends on customers without direct interaction with providers.
2. **Broad Network Access:** Service must be available to any device using any network.
3. **Resource Pooling:** Provider creates a pool of resources and dynamically allocates them to customers.

4. **Rapid Elasticity:** The services provided by the provider must be easily expandable and quick.
5. **Measured Services:** Provider must measure the usage of service and charge it accordingly. Tracking usage is also helpful in improving services.

Below is a table of differences between Public Cloud and Private Cloud is as follows:

Public Cloud	Private Cloud
Cloud Computing infrastructure is shared with the public by service providers over the internet. It supports multiple customers i.e, enterprises.	Cloud Computing infrastructure is shared with private organizations by service providers over the internet. It supports one enterprise.
Multi-Tenancy i.e, Data of many enterprises are stored in a shared environment but are isolated. Data is shared as per rule, permission, and security.	Single Tenancy i.e, Data of a single enterprise is stored.
Cloud service provider provides all the possible services and hardware as the user-base is the world. Different people and organizations may need different services and hardware. Services provided must be versatile.	Specific services and hardware as per the need of the enterprise are available in a private cloud.
It is hosted at the Service Provider site.	It is hosted at the Service Provider site or enterprise.
It is connected to the public internet.	It only supports connectivity over the private network.
Scalability is very high, and reliability is moderate.	Scalability is limited, and reliability is very high.
Cloud service provider manages the cloud and customers use them.	Managed and used by a single enterprise.
It is cheaper than the private cloud.	It is costlier than the public cloud.
Security matters and dependent on the service provider.	It gives a high class of security.
Performance is low to medium.	Performance is high.
It has shared servers.	It has dedicated servers.
Example: Amazon web service (AWS) and Google AppEngine etc.	Example: Microsoft KVM, HP, Red Hat & VMWare etc.

Cloud Ecosystem

A *Cloud Ecosystem* refers to the interconnected network of cloud computing services, solutions, technologies, and providers that collectively enable the delivery, management, and consumption of cloud resources and services. It encompasses the entire cloud computing landscape, including infrastructure, platform, software, and other related components. Cloud ecosystems are designed to support various use cases, meet diverse business needs, and foster innovation within the cloud computing environment.

cloud ecosystem is a complex system of interdependent components that all work together to enable cloud services. In nature, an ecosystem is composed of living and nonliving things that are connected and work together. In cloud computing, the ecosystem consists of hardware and software as well as cloud customers, cloud engineers, consultants, integrators and partners.

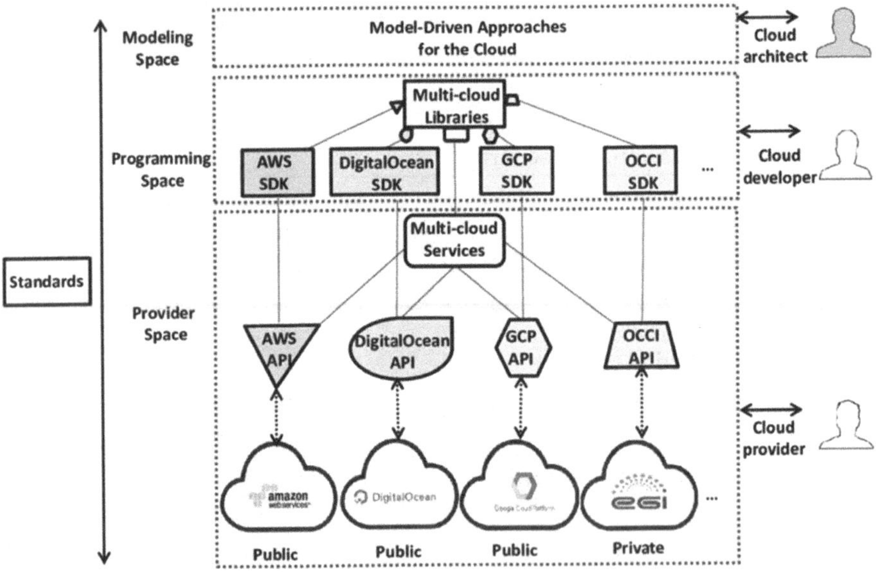

How a cloud ecosystem works

The center of a cloud ecosystem is a public cloud provider. It might be an IaaS provider such as Amazon Web Services (AWS) or a SaaS vendor such as Salesforce. Radiating out from the center of the cloud are software companies that use the provider's anchor platform, as well as consultants and companies that have formed strategic alliances with the anchor provider.

There is no vendor lock-in because these companies overlap, making the ecosystem more complex. For example, AWS is the center of its own ecosystem, but it's also a part of the Salesforce ecosystem. Salesforce runs a number of its services on AWS's infrastructure, and Salesforce customers can gain access, through devices called connectors, to pieces of AWS, such as its Simple Storage Service (S3).

A robust ecosystem provides a cloud provider's customers with an easy way to find and purchase business applications and respond to changing business needs. When the apps are sold through a provider's app store such as AWS Marketplace, Microsoft Azure Marketplace (for cloud software) or Microsoft AppSource (for business applications), the customer essentially has access to a catalog of different vendors' software and services that have already been vetted and reviewed for security, risk and cost.

How cloud ecosystems are used:

1. **Cloud Service Providers (CSPs):** CSPs like Amazon Web Services (AWS), Microsoft Azure, Google Cloud Platform (GCP), and IBM Cloud are central components of the cloud ecosystem. They offer a wide range of cloud services, including computing, storage, databases, analytics, and AI, which organizations can leverage to build and scale their IT infrastructure.
2. **SaaS Applications:** Software as a Service (SaaS) providers, such as Salesforce, Adobe Creative Cloud, and Dropbox, offer applications and services that are accessed by users over the internet. These applications are hosted in the cloud and are a key part of the cloud ecosystem for businesses and individuals.
3. **PaaS and Development Tools:** Platforms as a Service (PaaS) providers, like Heroku, Google App Engine, and Microsoft Azure App Service, offer development platforms and tools that streamline application development, deployment, and management in the cloud.
4. **DevOps and Continuous Integration/Continuous Deployment (CI/CD):** Tools and services for DevOps and CI/CD, such as Jenkins, Travis CI, and GitLab CI/CD, are integral to the cloud ecosystem. They enable organizations to automate software development processes, testing, and deployment in cloud environments.
5. **Container Orchestration:** Container orchestration platforms like Kubernetes play a vital role in managing and scaling containerized applications within the cloud ecosystem. They provide the infrastructure for deploying and managing containers efficiently.
6. **Serverless Computing:** Serverless platforms, such as AWS Lambda and Azure Functions, are part of the cloud ecosystem, allowing developers to run code in response to events without managing server infrastructure.
7. **Database Services:** Cloud providers offer a variety of database services, including relational databases (e.g., Amazon RDS, Azure SQL Database), NoSQL databases (e.g., Amazon DynamoDB, Azure Cosmos DB), and data warehousing (e.g., Google BigQuery), which are essential components of many cloud-based applications.

8. **IoT and Edge Computing:** Cloud ecosystems support IoT and edge computing by providing the infrastructure and services needed to collect, process, and analyze data from IoT devices and edge locations. AWS IoT and Azure IoT are examples of cloud services in this domain.
9. **AI and Machine Learning:** Cloud providers offer AI and machine learning services, such as Amazon SageMaker, Azure Machine Learning, and Google AI Platform, which enable organizations to develop and deploy machine learning models in the cloud.
10. **Data Analytics and Big Data:** Cloud ecosystems offer data analytics and big data services like Amazon EMR, Azure HDInsight, and Google BigQuery to help organizations analyze large datasets and gain insights from their data.
11. **Content Delivery and CDN:** Content delivery networks (CDNs) like Amazon CloudFront and Akamai are part of the cloud ecosystem. They help deliver web content, videos, and other media to users around the world with low latency and high performance.
12. **Hybrid and Multi-Cloud Solutions:** Solutions that facilitate hybrid and multi-cloud deployments, such as VMware Cloud on AWS and Anthos by Google Cloud, are integral to the cloud ecosystem. They enable organizations to integrate on-premises infrastructure with cloud resources.
13. **Cloud Security and Identity Management:** Security services and identity management solutions, like AWS Identity and Access Management (IAM) and Azure Active Directory, are essential components that help organizations secure their cloud environments and manage user access.
14. **Marketplaces and Third-Party Integrations:** Many cloud ecosystems include marketplaces where third-party vendors offer complementary services, applications, and integrations that enhance the capabilities of the cloud platform. For example, the AWS Marketplace and Azure Marketplace provide a wide range of third-party solutions.

The cloud ecosystem is dynamic and ever-expanding, with continuous innovation and the introduction of new services and technologies. Organizations leverage the components of the cloud ecosystem to build, scale, and optimize their cloud-based solutions, addressing a wide range of business needs and requirements.

On-Demand Computing

On-demand computing in cloud computing refers to the ability to quickly and easily access computing resources and services whenever they are needed, without requiring manual intervention or long-term commitments. It allows users to provision and utilize computing resources on a pay-as-you-go basis, aligning costs with actual usage.

With on-demand computing, users can request and obtain resources from the cloud provider in a self-service manner, usually through web-based interfaces or APIs. The cloud provider dynamically allocates and provisions the requested resources, such as virtual machines, storage, and networking, based on the user's specifications.

Key aspects of on-demand computing include:

1. **Resource Provisioning:** Users can request computing resources on-demand, without needing to own or maintain the underlying physical infrastructure. The cloud provider automatically allocates and provisions the requested resources based on availability and the user's requirements.
2. **Instant Availability:** On-demand computing enables quick access to computing resources. Users can obtain the resources within minutes or even seconds, allowing them to rapidly scale up or down based on their needs. This agility is especially beneficial during peak periods or sudden spikes in demand.
3. **Flexibility and Scalability:** On-demand computing allows users to easily scale their resources up or down as required. They can increase resource capacity during periods of high demand and decrease it when demand subsides. This scalability ensures optimal resource utilization and cost-efficiency.
4. **Pay-as-you-go Model:** With on-demand computing, users pay only for the resources they consume. Costs are typically based on usage metrics such as the number of virtual machine hours, storage capacity, network bandwidth, or data transfer. This pay-as-you-go model eliminates the need for upfront investments and provides cost transparency and control.
5. **Self-Service Provisioning:** On-demand computing emphasizes self-service, empowering users to provision resources autonomously. Users can select the desired resource configurations, specify the quantity, and initiate the provisioning process themselves. This self-service capability reduces administrative overhead and enables faster deployment of applications and services.
6. **Dynamic Allocation:** On-demand computing relies on the cloud provider's ability to dynamically allocate and de-allocate resources based on user demand. This dynamic allocation ensures efficient resource utilization across multiple users and allows the provider to optimize their infrastructure for maximum efficiency.

On-demand computing is a fundamental concept in cloud computing and is a key enabler of its benefits, such as cost optimization, scalability, agility, and rapid deployment. It allows businesses to respond quickly to changing demands, experiment with new ideas without significant upfront investments, and efficiently utilize computing resources on a flexible and as-needed basis.

Introduction to SaaS in Web Development

Software as a Service (SaaS) has revolutionized the way businesses and developers approach web development. SaaS refers to cloud-based software tools and platforms used to build, manage, and deploy web applications, eliminating the need for traditional on-premises software installation and maintenance. These cloud-based solutions provide a more accessible, streamlined, and cost-effective way for developers and businesses to create, host, and maintain web-based applications.

With the rise of SaaS in web development, various tools and platforms have emerged, catering to diverse levels of technical proficiency and project demands. SaaS tools empower developers to construct sophisticated web applications while concentrating on enhancing the core functionality and user experience without being burdened by intricate infrastructure setup and maintenance. As a result, both small businesses and large enterprises have embraced SaaS-based web development as their preferred approach for building and expanding web applications.

Benefits of SaaS Solutions for Web Development

There are several advantages to using SaaS solutions for web development, which have made them a popular choice among businesses and developers.

- **Cost reduction:** SaaS solutions typically have a subscription-based pricing model, which means businesses can pay for only the features and resources they use, rather than investing in expensive hardware, licenses, and infrastructure. This approach reduces upfront costs, making it more affordable for small businesses to build web applications.
- **Scalability:** SaaS platforms offer the flexibility to scale resources up or down based on user demand and requirements. As your web application grows, you can easily add more resources and capabilities without significant downtime or expensive infrastructure upgrades.
- **Ease of use:** Many SaaS web development tools offer intuitive drag-and-drop interfaces, simplifying the application development process for technical and non-technical users. This allows businesses to build web applications quickly, even without a team of expert developers.
- **Seamless updates:** With SaaS, updates and new features are automatically rolled out to users, ensuring they can access the latest software version without any manual intervention. This reduces the time and effort spent on software updates, allowing developers to focus on improving their web applications.
- **Increased collaboration:** SaaS platforms typically offer cloud-based collaboration tools that enable teams to work together more effectively during the web development process. This facilitates real-time communication, file sharing, and feedback, leading to more efficient workflows and a faster project turnaround.

- **Faster time to market:** Easy-to-use tools, reduced infrastructure requirements, and collaborative features allow for a quicker web development process. This results in a faster time to market for new web applications, giving businesses a competitive edge in the digital industry.

Popular SaaS Web Development Tools and Platforms

The widespread adoption of SaaS in web development has led to various tools and platforms catering to various needs and requirements. Here are some popular SaaS web development tools and platforms that you might consider for your projects.

1. **Shopify** is an all-in-one e-commerce platform allowing businesses to set up online stores and sell products. It offers a wide range of customizable templates, payment integrations, and marketing tools, making it a popular choice for small businesses and entrepreneurs looking to establish an online presence.
2. **Wix** is a website builder with an easy-to-use drag-and-drop editor, allowing users to create responsive and visually appealing websites without writing any code. With its extensive template library and app market, Wix is an excellent SaaS solution for those looking to create a website quickly and easily.
3. **Squarespace** is another famous website builder that focuses on providing visually appealing, responsive templates for various industries. It offers a range of built-in features, such as e-commerce, blogging, and booking capabilities, making it an ideal choice for businesses that want an attractive website with minimal effort.
4. **WordPress.com** is different from open-source WordPress.org. It's a hosted platform that provides various tools for creating and managing websites, such as customizable themes, plugins, and built-in SEO tools. They offer free and premium plans, making it suitable for different users, including bloggers, small businesses, and e-commerce sites.
5. **AppMaster.io** is a powerful no-code platform that helps users create backend, web, and mobile applications visually. Its tools let you design data models, build business logic through visual Business Process Designers, and create responsive UIs with a drag-and-drop interface. AppMaster.io generates source code and compiles applications, ensuring reduced technical debt and high-performance web applications. Its scalable and cost-effective approach makes it a popular choice among businesses and developers of all sizes.

FREQUENTLY ASKED QUESTIONS

1. Enlist any two characteristics of cloud computing.
2. Differentiate between Public cloud and Private cloud.(any two points)
3. State features of Cloud Service Models (any four points).
4. List any four general issues in securities of cloud computing.
5. Define On-Demand Computing
6. Explain cloud computing architecture.
7. Differentiate Public and Private cloud.
8. Classify the cloud computing services.
9. How cloud provides Infrastructure as a service (IAAS)? Explain.
10. What does the acronym SaaS mean? How does it relate to cloud computing?
11. Identify the Web-Based Communication Tools used in SaaS.
12. List and discuss Security implications of Cloud Computing.
13. Define Cloud Computing.
14. Identify any two advantages of distributed computing.
15. How is On Demand provisioning of resources applied in cloud computing?